Current Topics in Soil Science

Swapna Mukherjee

Current Topics in Soil Science

An Environmental Approach

Springer

Swapna Mukherjee
Formerly at Geological Survey of India
Kolkata, India

Jointly published with Capital Publishing Company, New Delhi, India

The print edition is not for sale in Afghanistan, Bangladesh, Bhutan, India, the Maldives, Nepal, Pakistan and Sri Lanka. Customers from Afghanistan, Bangladesh, Bhutan, India, the Maldives, Nepal, Pakistan and Sri Lanka please order the print book from: Capital Publishing Company, 7/28, Mahaveer Street, Ansari Road, Daryaganj, New Delhi 110 002, India.

Every effort has been made to contact the copyright holders of the figures and tables which have been reproduced from other sources. Anyone who has not been properly credited is requested to contact the publishers, so that due acknowledgment may be made in subsequent editions.

ISBN 978-3-030-92668-7 ISBN 978-3-030-92669-4 (eBook)
https://doi.org/10.1007/978-3-030-92669-4

This Springer imprint is published by the registered company Springer Nature Switzerland AG
The registered company address is: Gewerbestrasse 11, 6330 Cham, Switzerland

Preface

Soil science is a multidisciplinary subject introduced in the 1800's. It is the study of classification and formation of soil; its physical, chemical and biological properties and their use in soil management. These are discussed briefly in the first four parts of the book. The world is now much more concerned about the environment than before as it is declining at a much faster rate. The impact is now more obvious and is affecting all aspects of daily lives. Thus, apart from some introduction to soil science, this book is much more focused on the recent environmental impacts on soil and mitigation practices. This is because soil is needed to be understood first before it is improved.

There are a number of well referred books of soil science viz. Phogat, 2015; Foth, 1990; Sumner, 1999; Lal, 2007. This book is meant for students who are introduced to soil science in graduation and post-graduation levels. The book gives a general understanding of the subject and will help specially motivated students from different disciplines to have a platform for higher studies in soil science. Emphasis is given in the descriptive part rather than mathematical part with few exceptions in soil physics. Most of the descriptions are accompanied by figures and diagrams to make things easily understandable.

The book starts with an introduction of soil science and the importance of their study. It is divided into five parts.

This book begins with the basic concepts of soil science including soil formation, mineralogy and classification. This portion also briefly describes the concept of soil quality and various analytical techniques needed to study the properties of soil. It is very important to understand the techniques which are needed to study the respective properties of soil.

Next comes basic concepts of soil physics e.g. various physical properties of soil such as its size, shape, surface area, density etc. Soil structure describes the classification and genesis of various types of structures in soil. Soil tillage explains its types and some modern tillage practices. They also inform some of the effects on soil quality with some case studies. Next chapters include various concepts of interactions of soil with water, air and temperature. The last chapter of this part describes soil erosion which includes water and wind erosion. This chapter states how erosion is affected by various factors and how it can be controlled.

Soil chemistry is described and discussed in the next phase. This includes basic concepts in soil chemistry and chemical composition of soil. Other chapters include brief description of various chemical properties of soil such as ion exchange, pH, salinity, sodicity etc. The last chapter discusses the importance of macronutrients and micronutrients and their functions in soil.

Soil biology starts with soil microbiota and soil fauna and their interaction in soil. This book also discusses the positive and negative effect of pesticides and fertilizers on microbial diversity of soil.

Finally, with an introduction of various concepts of soil environment such as green manure, cover crops, waste management etc. it also discusses various environmental concerns and their causes and impact on soil. The next few chapters broadly mention and discuss these environmental concerns and how they can be mitigated. Environmental concerns include climate change, soil pollution etc. and mitigating processes such as carbon sequestration, nutrient and fertility management.

Formerly at Geological Survey of India
Kolkata, India

Swapna Mukherjee

Acknowledgement

The completion of the book resulted from the cordial help and encouragement received from the soil scientists of Kolkata and Delhi. I would like to acknowledge them and in particular, Prof. S. K. Sanyal, ex. VC of Kalyani University, Dr. Krishnendu Das of ICAR, Kolkata, INSA member, Dr. Kunal Ghosh of Kolkata and my friends Dr. S. Datta, Dr. N. Ahmed, Dr. S. Mahapatra and Dr. S. Mukhopadhayay of the Clay Society of India. In particular I am indebted to Dr. Stephan Peth of the Soil Science Department of Kassel University, Germany for his immense help for providing me very useful material related to soil science.

I deeply appreciate the immense help of my technical assistant Ms Amrita Mukherjee, a young and bright student of Geology, Calcutta University, who competently helped in preparing the manuscript. I acknowledge the help of many of my friends in soil science across India for all the useful discussions we had. The utility and success of the book will depend on the benefit the student community can derive from it and for whom it has been written.

Lastly, I deeply appreciate the constant encouragement and support of my husband Prof. P.K. Mukherjee.

The book is dedicated to our daughter Ms Swagata Mukherjee.

Introduction

Soil is a natural body that is engaged in dynamic interactions with the atmosphere and lithosphere influencing the planet's climate and hydrological cycle and serving as a growth medium for a versatile community of living organisms. It also plays a major environmental role as a biological, physical and chemical reactor that decomposes waste products and recycles them into nutrients for the continual regeneration of life on earth.

Soil science is a study of soils that deals with soil as a natural resource on the surface of the earth including soil formation, classification and its minerological composition. Physical, chemical, biological, and fertility properties of soils is required in relation to the use and management of soils.

Branches of soil science includes:

- Pedology, that deals with formation, genesis, and description of soil properties which has been concluded in Soil Physics.
- Edaphology, that studies soil in relation to growth, nutrition and yield of crops or plants which has been concluded in Soil Chemistry and Soil Biology.

The main concern of this book is soil environment and the current issues being faced. For this, knowledge of different branches of soil science is needed and has been discussed thoroughly.

Soil as an element has always been neglected in the climate system. People had always been concerned with air and water pollution and had ignored the largest carbon sink after the oceans.

It came to notice in 18th century when large number of crops started to fail in Europe. Large number of scientist started to study the processes controlling soil fertility. It paved the way for 19th century study in mineral fertilizers that saw a boom in agricultural productivity. Again, in the 1930s, public concern over the health and environmental impacts of the major dust storms associated with the "Dust Bowl" in the United States caused a significant research effort and major soil conservation programmes to be launched (Baveye et al., 2011; Sylvester and Rupley, 2012).

People started understanding that the quality of soil is declining over time and is causing huge impact in human life and the very need to conserve it for future generations. Currently, humanity is facing two long term major threats related to soil, viz., food security and ecological threats brought about by global climate change. And for that increasing land area for agriculture or plumbing more water from ground for irrigation is not an option. Instead, we need to modify agricultural practices and focus on natural nutrients to increase productivity.

There is no doubt that current climate change is having huge impact on soil. Current estimates of global soil C are slightly over 4000 Pg C (Petagrams of carbon), which is more than five and a half times the amount of carbon currently in the atmosphere. The increasing concentration of carbon dioxide in our atmosphere may cause the microbes in the soil to work faster to break down organic matter, potentially releasing even more carbon dioxide. There are already signs that soil moisture content is being affected by rising temperatures and changes in precipitation patterns and future projections show this may continue.

However, with improving technologies it is more and more evident that the carbon content in soil is decreasing and studying the microbial activity of soil is getting very difficult with the fast changing environment. Thus, the need for measurement of structure is to be done in microscopic level rather than on spatial level. For many years, the only equipment that was of any help in this respect was the scanning electron microscope, which could be used to visualize microbial micro-environments in soils (Foster, 1988), but unfortunately delivered no quantitative or spatial data. This situation has changed dramatically in the last few years. Significant technological advances have provided soil researchers with routine access to X-ray computed tomography systems, which, as methodological roadblocks are being resolved (Elliot and Heck, 2007; Iassonov et al., 2009; Baveye et al., 2010; Iassonov and Tuller, 2010; Hapca et al., 2013; Houston et al., 2013a,b), increasingly provide reliable information about the geometry of pores and solids in soils at resolutions as small as 0.5 μm.

The main concern of this book is soil environment and the current issues we are facing right now. For this, knowledge of different branches of soil science i.e. soil physics, chemistry and biology is needed which has been discussed thoroughly in this book. Some of the current researches in these fields are discussed below.

Soil Physics

As soil scientists, their common goal is to reduce agrochemical applications and maximize agricultural production without soil and water degradation. One of the important focus of soil physics is the subsurface water flow and chemical transport that may lead towards many environmental issues which include soil contamination and groundwater pollution by pesticides and fertilizers. Chemicals migrating from industrial and municipal disposal sites are creating similar problems to the environment.

Impact of tillage on soil physical properties and quality is also a growing concern among researchers. Compared to conventional tillage, no-tillage improves soil water storage capacity, bulk density and root system. These increase the crop yield to a significant amount. Studies on decreasing the rate of N fertilizer usage with no-tillage on wheat and maize has shown increase in soil C and N stocks. Soil scientists aim to study different soil management practices with different fertilization rates. This is to come up with the best solution to decrease environmental damage and improve soil quality.

Major soil threats due to human activities impair the capacity of soil to carry out important functions. These threats include soil erosion, compaction, sealing, floods and landslides. Overexploitation, overgrazing, inappropriate clearing techniques, and land use mismanagement have resulted in severe nutrient decline, water and wind erosion, compaction, and salinization. This has decreased the soil quality and subsequently soil productivity. In Africa, India, and China, erosion rates are 30-40 times faster than the rate of replenishment. Appropriate measures such as increasing plant cover, mulching, use of cover crops have been undertaken worldwide to replenish degraded soil.

Soil Chemistry

Soil is the main source of human nutrition. Thus, terrestrial plants remain the cheapest and efficient source of nutrition to mankind. Early researches were mainly designed to understand the chemical processes in soil so that growth of crop can be maximized. But recent researches revolves around more environmental friendly approaches for healthier food production and minimizing impacts on ecosystem.

Excessive use of fertilizers and pesticides can modify the soil chemistry more than the actual requirements. This can lead to toxicity in soil which not only lead to decrease in soil crop health but can also affect animals consuming it. Recently importance has been given to researches concerning behaviour of contaminants such as heavy metals at the soil-water interface. Thus, proper management practices are required for protection of food crops from toxic metal contaminants. For example, arsenic accumulation in rice is a global concern. Other common heavy metals found in food crops are mercury, cadmium, lead and chromium. These contaminants are found in crops like lettuce, amaranth, water spinach, cowpea, and rice.

Similarly, less application of some nutrients leads to crop deficiency. Minimizing trace element deficiencies is one of the most important concerns among soil scientists. Zn deficiency is seen in both human health and crop production. Thus researches concerning processes to improve movement of Zn from soil to edible crops have been studied. Micronutrients such as manganese and copper have been seen to be deficient in many plants.

Anthropogenic activities from mining to simple salting on roadside can change the chemistry of soil. Mining can lead to high metal concentration in soil whereas excessive salting can disrupt the nitrogen cycle. Livestock manure, irrigation with wastewater, metallo-pesticides, phosphate-based fertilizers and sewage sludge–based amendments can also be the primary sources of heavy metal contamination in plants. Thus it is very important to understand the amount of essential nutrients needed by plants.

Soil Biology

Soil scientists are most interested in microbial diversity in soil and their relationship between them. Each organism is studied very closely to understand their role in soil environment. For example, recent studies on Protists have shown that they play a major role in enhancing plant growth and act as bioindicators. They also act as agricultural pests and regulate population of other communities.

Soil biology is also studied in relation to soil chemistry and physics. In soil physics, soil biology is used to predict emission of greenhouse gases from soil. Models on gaseous transport and the biochemical processes in the soil are produced for this study. This is because CH_4 and N_2O are both produced and consumed in soil. The eventual flux to the atmosphere depends on the position of reaction sites and the escape pathways for these gases.

The impacts of climate change on microbes and their extracellular enzymes are not well understood by scientists. Numerous studies have been done to predict, assess and manage the activities of these enzymes and their importance in carbon sequestration and plant growth.

Recently, shifting to sustainable agriculture has forced farmers to opt for more ecofriendly techniques to improve soil quality. Increased use of herbicides has shown to effect biological activities in soil. These include disruptions to earthworm ecology in soils, inhibition of soil N-cycling in alkaline or low organic matter soils; and site-specific increase in disease resulting from the application of a variety of herbicides. A recent study conducted on rice-wheat-maize crop rotations with a combination of tillage practices has shown that soil biodiversity has improved drastically with conservation tillage practices. A combination of zero tillage and mulching showed highest microbial population (bacteria, fungi and actinomycetes).

Importance of soil is increasing and so is its threats like erosion, urban development, pollution, climate change etc. This is prompting soil scientists to consider the application of concepts of endangered species to soil. These are named rare or threatened soil around the world. Efforts have been made to identify and map these soils with the help of GIS sensors. For example in 2014, China released their first pedodiversity survey where they identified 90 endangered soils and 24 type of soils that have gone extinct. Similar surveys have also been made in Russia, Europe and USA. Thus efforts are being made by soil science communities to pressurise the government on creating reserves for these soils just like endangered plants and animals.

References

Abdel-Raouf, N., Al-Homaidan, A.A. and Ibraheem, I.B.M. (2012). Agricultural importance of algae. *African Journal of Biotechnology*, **11:** 11648–11658.

Aide, M., Beighley, D. and Dunn, D. (2016). Arsenic in the soil environment: A soil chemistry review. *International Journal of Applied Agricultural Research*, Research India Publications, **11:** 1–28.

Blagodatsky, S. and Smith, P. (2011). Soil physics meets soil biology: Towards better mechanistic prediction of greenhouse gas emissions from soil. *Soil Biology and Biochemistry*, Elsevier, **47:** 78–92.

Burns, R.G., DeForest, J.L., Marxsen, J., Sinsabaugh, R.L., Stromberger, M.E., Wallenstein, M.D., Weintraub, M.N. and Zoppini, A. (2013). Soil enzymes in a changing environment: Current knowledge and future directions. *Soil Biology and Biochemistry*, Elsevier, **58:** 216–234.

Castro-Huerta, R.A., Falco, L.B., Sandler, R.V. and Coviella, C.E. (2015). Differential Contribution of Soil Biota Groups to Plant Litter Decomposition as Mediated by Soil Use. *Peer J.*, DOI:10.7717/peerj.826.

Choudhary, M., Datta, A., Jat, H.S., Yadav, A.K., Gathala, M.K., Sapkota, T.B., Das, A.K., Sharma, P.C., Jat, M.L., Singh, R. and Ladha, J.K. (2018). Changes in soil biology under conservation agriculture based sustainable intensification of cereal systems in Indo-Gangetic Plains, *Geoderma*, Elsevier, **313:** 193–204.

Coleman, D.C. and Wall, D.H. (2015). Soil Fauna: Occurrence, Biodiversity, and Roles in Ecosystem Function. Elsevier Inc.

Doula, M.K. and Sarris, A. (2016). Soil Environment. Elsevier B.V.

Gao-bao, H., Qiang, C., Fu-xue, F. and Ai-zhong, Y. (2012). Effects of different tillage systems on soil properties, root growth, grain yield, and water use efficiency of winter wheat (*Triticum aestivum* L.) in arid Northwest China. *Journal of Integrative Agriculture*, **11:** 1286–1296.

Geisen, S., Mitchell, E.A.D., Adl, S., Bonkowski, M., Dunthorn, M., Ekelund, F., Fernández, L.D., Jousset, A., Krashevska, V., Singer, D., Spiegel, F.W., Walochnik, J. and Lara, E. (2018). Soil protists: A fertile frontier in soil biology research. *Microbiology Reviews*, **42:** 293–323.

Green, S.M., Machin, R. and Cresser, M.S. (2008). Effect of long-term changes in soil chemistry induced by road salt applications on N-transformations in roadside soils. *Environmental Pollution*, Elsevier, **152:** 20–31.

Impa, S.M. and Johnson-Beebout, S.E. (2012). Mitigating zinc deficiency and achieving high grain Zn in rice through integration of soil chemistry and plant physiology research. *Plant Soil*, **361:** 3–41.

Liu, Z., Sun, K., Liu, W., Gao, T., Li, G., Han, H., Li, Z. and Ning, T. (2019). Responses of soil carbon, nitrogen, and wheat and maize productivity to 10 years of decreased nitrogen fertilizer under contrasting tillage systems. S*oil and Tillage Research*, Elsevier, **196:** 1–11.

Nkongolo, K.K., Spiers, G., Beckett, P., Narendrula, R., Theriault, G., Tran, A. and Kalubi, K.N. (2013). Long-term effects of liming on soil chemistry in stable and eroded upland areas in a mining region. *Water, Air, and Soil Pollution,* Springer, **224**, DOI: https://doi.org/10.1007/s11270-013-1618-x.

Osman, K.T. (2014). Soil Degradation, Conservation and Remediation. Springer Science+Business Media Dordrecht.

Pareja-Sánchez, E., Cantero-Martínez, C. and Álvaro-Fuentes, J. (2020). Impact of tillage and N fertilization rate on soil N_2O emissions in irrigated maize in a Mediterranean agroecosystem. *Agriculture, Ecosystems and Environment*, Elsevier B.V, **287:** 1–10.

Phogat, V.K., Tomar, V.S. and Dahiya, R. (2015). Soil Physical Properties. Soil Science: An Introduction. Indian Society of Soil Science.

Rai, P.K., Lee, S.S., Zhang, M., Tsang, Y.F. and Kim, K.H. (2019). Heavy metals in food crops: Health risks, fate, mechanisms, and management. *Environmental International*, Elsevier, **125:** 365–385.

Raiesi, F. and Kabiri, V. (2016). Identification of soil quality indicators for assessing the effect of different tillage practices through a soil quality index in a semi-arid environment. *Ecological Indicators*, Elsevier, **71:** 197–207.

Rose, M.T., Cavagnaro, T.R., Scanlan, C.A., Rose, T.J., Vancov, T., Kimber, S., Kennedy, I.R., Kookana, R.S. and Zwieten, L.V. (2016). Impact of Herbicides on Soil Biology and Function, Advances in Agronomy. Elsiever.

Strawn, D.G., Bohn, H.L. and O'Connor, G.A. (2020). Soil Chemistry. John Wiley and Sons Ltd.

Tennesen, M. (2014). Rare Earth. *American Association for the Advancement of Science*, **346:** 692–695.

Contents

Part III Soil Chemistry

Part I
Soil Science

Chapter 1
Soil Formation

Soil acts as an interface between atmosphere and lithosphere forming a thin mantle over the earth surface. It is a multiphase system consisting a mixture of water, gas, mineral matter and organic material. It also consists of soil organisms that uses soil not only as their habitat but also as a source of energy. This and various other factors changes the physical and chemical properties of soil that makes each soil unique. To understand formation of soil, we have to understand the pedogenic processes and the factors that lead to changes in its properties.

SOIL FORMING PROCESSES

For the formation of soil and its development, two main processes are involved. First formation of parent material and second, the development of soil layers. Many people confuse bedrock as parent material but the unconsolidated debris that are formed from the bedrock by weathering act as the parent material for soil. Rate at which parent materials are formed depends on many factors like climate, type of mineral etc. which will be discussed.

Weathering can be both physical, chemical and biological.

- **Physical weathering** occurs through the action of water, wind, and changes in temperature and pressure, which progressively break down rock into finer particles. In mountainous region, repeated action of freeze-thaw of water can break rocks apart. In deserts, vast temperature difference in day and night causes rocks to expand and contract. This causes surfaces of rocks to peel of like onion (Fig. 1.1).
- **Chemical weathering** involves the decomposition of minerals by a range of processes including solution, hydration, oxidation, and hydrolysis. For example, acidic water in contact with carbonate rocks can dissolve them to form limestone caves and pavements.

S. Mukherjee, *Current Topics in Soil Science*,
https://doi.org/10.1007/978-3-030-92669-4_1

Fig. 1.1 Onion skin weathering or exfoliation due to vast changes of temperature in day and night.

- **Biological weathering** occurs under the influence of organisms and plants. Release of organic acids by plants can decompose minerals causing chemical weathering or plant roots penetrating into rocks to disintegrate them (physical weathering).

Leaching is also an important pedogenic process in which downward movement of materials in soil solution takes place usually from one soil horizon to another. Element's solubility in water, effect of pH on that solubility, and the rate of water percolation through the soil influences its mobility.

SOIL PROFILE

Layers in soil are called soil horizons or strata and the arrangement of these horizons are called soil profile (Fig. 1.2). There is slight difference in strata and horizon. A strata may result from the pattern of deposition, or sedimentation, whereas when strata form in place by internal soil-forming (pedogenic) processes, they are called horizons. Each horizon in soil profile differ in properties such as colour, texture, structure, thickness, mineral content etc.

There are mainly three horizons in a soil profile, viz. A, B and C. Others are O, E and R.

- **Horizon O** is the uppermost horizon and is usually thin and dark due to the presence of humus. Upper part of the layer, mostly undecomposed is mostly from

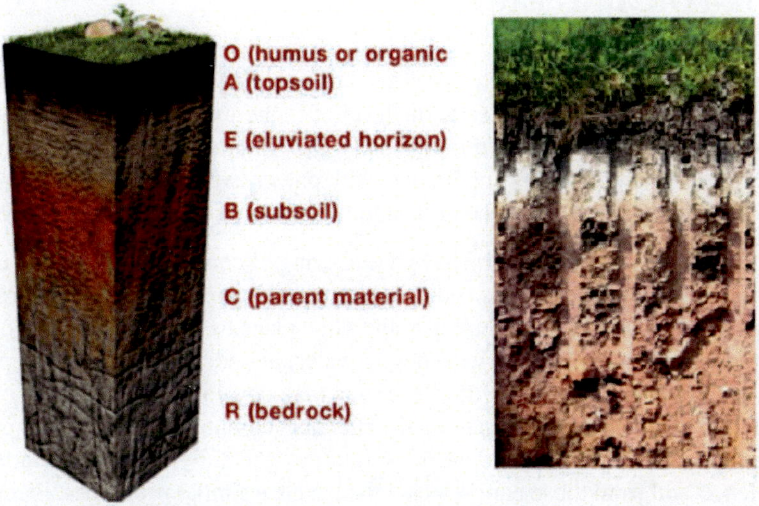

Fig. 1.2 Schematic representation of hypothetical soil profile (left) showing major horizons. Actual representation of soil profile (right).

leaf fragments whereas the lower O horizon contain organic debris in various stages of decomposition. Not all soils contain O horizon. Soils rich in organic matter ($\geq 20\%$) have this distinct layer.

- **Horizon A** often called the topsoil, this zone has the major biological activity and is therefore generally enriched with organic matter and typically darker in colour than the underlying soil. Here, plants and soil organisms (bacteria, fungi, earthworms etc) interact with each other for nutrients and energy. This horizon is generally soft and porous.
- **Horizon E** stands for eluviated and is pale in colour because of leaching of minerals and organic matter leaving mostly silica. This layer is present mostly in older or well-developed soil like forest soil.
- **Horizon B** is also called subsoil. B horizon generally consists of clay and mineral deposits more than upper layers due to leaching by percolating waters and accumulating in this layer. It is usually brownish or reddish in colour due to alteration of minerals to iron or aluminium oxide. The clay accumulation and the pressure of the overlying soil reduces the porosity of this layer. In some cases, an overly dense B horizon may inhibit aeration, internal drainage of water, and the penetration of roots.
- **Horizon C** is mainly formed from weathering of residual bedrock. C horizon is little effected by pedogenesis and so little or no clay is present from above layers. In transported soils C horizon may consist of alluvial, aeolian, or glacial deposits.
- **Horizon R** is partially weathered or unweathered bedrock at the base of the soil profile. It is a continuous mass of hard rock and is difficult to excavate by hand.

SOIL-FORMING FACTORS

The soil-forming factors are central to understanding the uniqueness in soils not only at landscape level but also at the level of individual soil profile. Although soil forming factors are considered individually, they interact with each other in nature, usually with a hierarchy of importance, with one or two of them being pre-eminent in soil development at a particular location. The main soil forming factors are:

- **Parent Material –** It is formed by weathering of consolidated bedrock in situ or from unconsolidated deposits derived from erosion of consolidated bedrock. The weathered products may remain in the same place to form residual soil or can be transported (by wind, glacier, running water) as sediments at different places to form deposits from which different soils may develop. Minerological composition of parent material influences the chemical composition of soil. For example, parent materials of basic igneous origin are generally rich in K, Ca and Mg. Hence, soil from those parent rock will also have similar properties. Parent material also influences grain size which in-turn determines soil texture, porosity, permeability, and cation exchange capacity (CEC). For example, soils formed from acidic lava such as granite and rhyolite are coarse textured and hence freely drained, with low cation retention capacity. The soils that typically develop are, therefore, strongly leached and nutrient-poor.
- **Climate –** Temperature and precipitation are the two most important climatic factors in soil development as they influence chemical weathering and decomposition of organic matter. Soil organisms are highly dependent on temperature. It is generally accepted that there is an approximate doubling of microbial activity and enzyme-catalysed reaction rates in soil for each 10 °C rise in temperature, upto around 30–35 °C (Bardgett, 2005). Some microorganisms can also live in extreme temperatures (pychrophiles and thermophiles). Altitude also increases soil organic matter (SOM). With increasing altitude, temperature decreases and precipitation increases thereby reducing microbial activity and rates of decomposition. It sometimes creates anaerobic conditions to form peat. Dramatic changes in plants takes place with both change in latitude and altitude thereby influencing the nitrogen content in soil. Too much precipitation can cause leaching resulting in acidic soils whereas very less precipitation causes accumulation of salts and carbonates forming saline soils.
- **Organisms –** Plants, animals and microorganisms affect soil formation. Soil biota can speed up or slow down the soil forming processes. For example, animals and microorganisms mix soils and form burrows and pores thereby destroying the soil horizons. Plants add organic matter to soil by fixing carbon from the atmosphere. Microorganisms like fungi, algae and bacteria develop soil by freeing plant nutrients bound in the organic matter and the rocks into forms more easily taken up by plants gradually increasing its fertility. Symbiotic relationship by plants and organisms causes nitrogen fixation.
- **Topography –** Topography influences soil development largely due to erosion and soil drainage which, depends on position of soil in a slope. Soils at or near

the top of a slope, tend to be freely drained with a water table at some depth, whereas those at or near the bottom of the valley, tend to be poorly drained with a water table close to the soil surface. Well drained soil tends to leach forming reddish layers rich in ferric iron oxides indicating oxidation whereas poorly drained soils tends to be anaerobic forming blue greyish layers of ferrous iron oxides indicating reduction.

- Steeper the slope, more the erosion. Thus the effect of soil erosion on steep slope makes the soil thickness narrow and coarse owing to the movement of finer soil particles down-slope. Erosion also changes the morphological features in soil such as forming terraces that causes soil creep.
- **Time –** It solely acts as a marker for the evolution of soil characteristics. With time and continuation of soil processes, soil becomes more matured and with increasing weathering it becomes more differentiated with thicker horizons. For example humus content in soil increases with time upto 10,000 years. After that soil loses its nutrients due to excessive leaching. With increasing organic content N content increases but P content decreases. This is due to the fact that as soils age, P becomes surrounded, or occluded, by Fe and Al hydrous oxides, rendering P largely unavailable to plants and also to soil biota.

PALEOSOL

Definition of paleosol in soil science is much different than that of geosciences. In soil science, paleosol refers to soils formed many years ago that have no relationship with the present day climate or vegetation due to differences in physical and chemical properties. Paleosol in geosciences mainly refers to fossil soils that are buried under sedimentary or volcanic deposits.

Paleosols older than 2.5 billion years have been found until now. With millions of years of climate change, properties of paleosol are far different than their surrounding environment. Thus they cannot be defined by soil taxonomy because high burial temperatures and pressures during diagenesis and lithification alters or destroys the colloidal fraction. Rather they have some criterias for identification:

- Formed in situ on bedrock.
- Soft-sediment deformation at the top of the soil profile.
- Up-profile changes in chemistry, texture, and mineralogy that is consistent with terrestrial weathering processes.

Since soils are formed in response to physical, chemical and biological changes, paleosols are used for paleoenvironment and paleoclimate reconstruction. Morphological features, clay minerology, stratigraphic changes are used as indicators for the reconstruction. For example, depth and width of dessication cracks can denote intensity of precipitation. Environments with greater rainfall and seasonality generate slickensides at greater depth (Vadivelu and Challa, 1985; Mermut et al., 1996a). In stratigraphical succession, vertical and horizontal variation in physical and chemical changes of paleosol give insights of environmental changes over time. For example carbonate formations such as calcite indicates water-stressed conditions in well-drained soils. Siderite in paleosol indicates suboxic to anoxic conditions, high concentrations of CO_2, and relatively low concentrations of Ca^{2+} and SO_4^{2-}, and hydromorphic soil conditions. Clay minerology are very good indicators of paleoclimate. Their significance are given in Table 1.1.

Recent paleosol research has introduced semi-quantitative and quantitative measures for chronometric reconstructions that provides information about changes in temperature, precipitation, and atmospheric pCO_2 (partial pressure of CO_2) both regionally and globally. The CO_2 levels in past atmosphere is reconstructed with the help of soil carbon gas exchange, carbon isotopes in pedogenic carbonate nodules and mass-balance approaches taking multiple atmospheric gases into account.

CONCLUSION

Soils evolve under the action of biological, climatic, geologic, and topographic influences. This causes soil profile to develop by forming distinctive soil horizons. Mainly five factors influence the soil formation, viz. parent material, climate, organisms, topography and time. Interaction of these factors causes soil formation. Indicators of paleosol such as morphology, clay minerals and stratigraphical sequence can give insight about the paleoenvironment and paleoclimate.

THINK FOR A WHILE

- Why bedrock are not considered parent materials?
- What are the properties in soil that influences leaching?
- What is the difference between a strata and a horizon?
- In which environment O Horizon is most prominent?

Table 1.1 Clay minerals in paleosol and their significance as paleoenvironment/paleoclimate indicators

Mineral	*Paleoenvironment/Paleoclimate indicators*
Vermiculite	Incipient chemical weathering (regional)
Kaolinite	Extensive chemical weathering (regional) promoted by high rainfall and leaching
Smectite	High base cation and Si activity in solution; promoted by seasonal rainfall
Goethite	Mildly acidic and oxidized soils, may undergo dehydration to hematite upon burial
Hematite	Alkaline and oxidized soils, may also be diagenetic
Gibbsite	Acidic soils characterized by intense chemical leaching
Illite	May form by postburial alteration of smectite-group clay minerals
Sepiolite	Alkaline pH and high Al activity in solution; arid to semiarid climate

- Why E Horizon is absent in many soils?
- What kind of soil will be formed from a basaltic bedrock?
- Why older soil has lesser phosphorus than younger soils?
- What insight siderite can give about the paleoenvironment with its presence in paleosol?

REFERENCES

Bardgett, R.D. (2005). The Biology of Soil: A Community and Ecosystem Approach. Oxford Press.

Foth, H.D. (1990). Fundamentals of Soil Science. John Wiley & Sons.

Hillel, D. (1998). Environmental Soil Physics. Academic Press.

Mermut, A.R., Dasog, G.S. and Dowuona, G.N. (1996a). Soil morphology. *Dev. Soil. Sci.*, **24**: 89–114.

Sumner, M.E. (2000). **Handbook of Soil Science. CRC Press.**

Tabor, N.J. and Myers, T.S. (2015). Paleosols as Indicators of Paleoenvironment and Paleoclimate, *Annual Reviews,* **43:** 333–361.

Vadivelu, S. and Challa, O. (1985). Depth of slickenside occurrence in Vertisols. *J. Indian Soc. Soil Sci.*, **33:** 452–454.

Chapter 2
Soil Minerology

Minerals are defined as solid, inorganic, naturally occurring chemical compounds with a definite chemical formula and general structure. Minerals found in soil is a result of weathering or the combined action of hydrosphere, atmosphere and biosphere in rock materials. Soil minerals can be formed from in situ or local rocks and are called authigenic. They can also be transported from a different parent rock distance away and are called detrital in origin.

Minerals can also be grouped as primary and secondary. Primary minerals are those which have not been altered chemically since their crystallization from molten lava, and secondary mineral results from the decomposition of a primary mineral or from the precipitation of the decomposition products of minerals. These minerals decompose or react with each other and release their elements to the soil solution.

Since soils develop from parent material composed of rocks and minerals from the earth's crust, knowledge in chemical and mineralogical composition of the earth's crust is very important (Table 2.1). Elements combine with each other to form minerals. Out of 100 elements found in earth's crust, around 2000 minerals have been recognized (Fig. 2.1). With respect to soil science, clay minerals play a pivotal role as the soil's microaggregate. Being the most reactive inorganic component in soil, clay minerals are explained in detail in this chapter.

The primary minerals identified in soils belong mainly to the classes of silicates, oxides of Fe, Zr, Ti and phosphates out of which 40% are silicates. The building unit of the silicates is the silicon tetrahedron in which one *silicon* atom is surrounded by and bonded to four oxygen atoms, each at the corner of a regular *tetrahedron*. Based on this different structures are found. Types of silicate structures are shown in Fig. 2.2 and their chemical formulas given in Table 2.2.

Since secondary soil minerals are formed from primary minerals during weathering, they are generally characterized by smaller particle size. They are the principal components of the silt and clay fraction of soils. The major secondary minerals are listed in Table 2.3.

S. Mukherjee, *Current Topics in Soil Science*,
https://doi.org/10.1007/978-3-030-92669-4_2

Table 2.1 Main elements in Earth's crust

	Crust Percent by	
Element (symbol)	*Mass*	*Volume*
Oxygen (O)	46.10	94.04
Silicon (Si)	28.20	0.88
Aluminium (Al)	8.23	0.48
Iron (Fe)	5.63	0.49
Calcium (Ca)	4.15	1.18
Sodium (Na)	2.36	1.11
Magnesium (Mg)	2.33	0.33
Potassium (K)	2.09	1.42

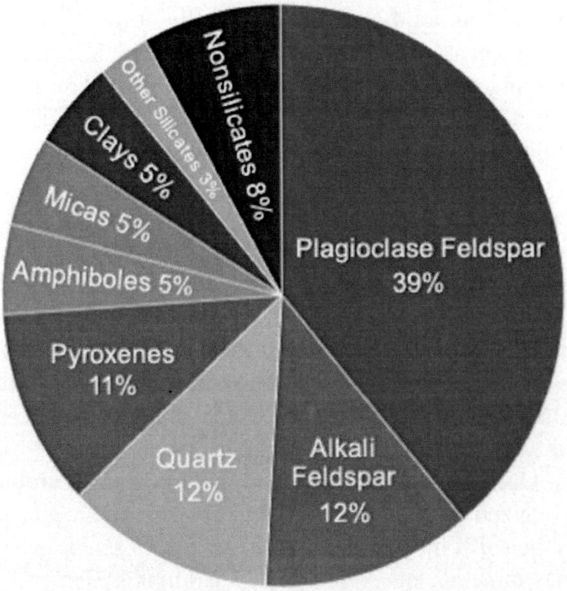

Fig. 2.1 Abundance of minerals in Earth's crust.

Clay Minerals

Clay minerals are a group of hydrous aluminosilicates that mainly occurs in the clay-sized (<2 μm) fraction of soil. They are mainly phyllosilicates in that they contain assemblages of tetrahedral and octahedral sheets. Tetrahedral sheets are an interlocking array of silica tetrahedrons connected at three corners in the same plane by shared oxygen anions to form a hexagonal network. When external ions bond to the tetrahedral sheet they are coordinated to one hydroxyl and two oxygen anion groups. An aluminum, magnesium, or iron ion typically serves as the coordinating cation and is surrounded by six oxygen atoms or hydroxyl groups resulting in an

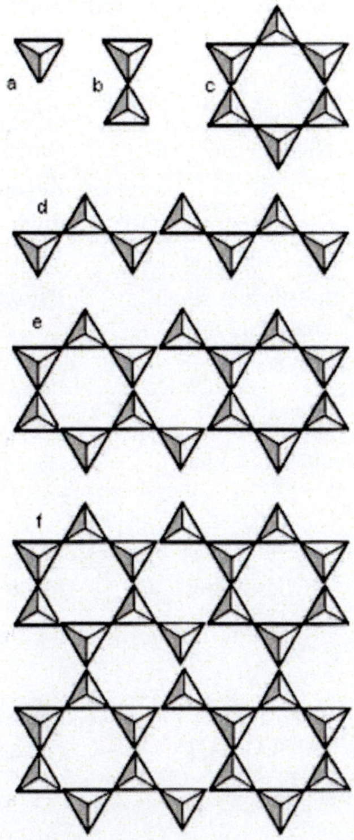

Fig. 2.2 Types of silicate structures: a. Nesosilicates, b. Sorosilicates, c. Cyclosilicates, d. Single chain silicates, e. Double chain silicates and f. Phyllosilicates.

eight-sided building block termed an octohedron. The horizontal linkage of multiple octahedra comprises the octahedral sheet (Barton and Karathanasis, 2002).

Cations in the octahedral layer may exist as divalent or trivalent. When the cations are divalent (Mg^{+2}, Fe^{+2}), the ratio of divalent cations to oxygen is 1:2 and all three possible cation sites in the octahedron are occupied. The octahedral sheet is called trioctahedra. When the cations are trivalent (Al^{+3}, Fe^{+3}), the charge balance is maintained by leaving one of every three octahedral cation sites empty. Under this configuration, the ratio of trivalent cations to oxygen is 1:3 and the layer is called dioctahedral sheet. Based on this combination of arrangements, different types of clay minerals formed are:

1:1 Clay Minerals – In this one tetrahedral sheet is bonded to one octahedral sheet. Ideal chemical formula is $Al_2Si_2O_5(OH)_4$. The dioctahedral version of 1:1 layer is the Kaolin group with Kaolinite being the most common mineral. Here silica tetrahedral sheet is bonded to an aluminium octahedral sheet by Van der Waals force. Other minerals in this group are dickite, nacrite, and halloysite. Serpentine group is the trioctahedral version of 1:1 layer and consists of minerals like antigorite and chrysotile.

2:1 Clay Minerals – In 2:1, two tetrahedral sheets are coordinated to one octahedral sheet. The ideal chemical formula would be $Al_2Si_4O_{10}(OH)_2$. The ideal group of this type is the pyrophyllite–talc group (Fig. 2.3). Each group is further subdi-

Table 2.2 Common primary minerals of the silicate class found in soils

Silicon tetrahedra arrangement	*Name of mineral or group*	*Ideal formula*
Nesosilicates $[SiO_4]^{4-}$	Olivine	$(Mg, Fe, Mn)_2SiO_4$
	Zircon	$ZrSiO_4$
Sorosilicates $[Si_2O_7]^{6-}$	Epidote	$Ca_2(Al,Fe)_3O(SiO_4)(Si_2O_7)(OH)$
Cyclosilicates $[Si_nO_{3n}]^{2n-}$	Tourmaline	$(Na,Ca)(Al,Li,Mg)_{3-}(Al,Fe,Mn)_6(Si_6O_{18})(BO_3)_3(OH)_4$
Inosilicates single chain $[Si_nO_{3n}]^{2n-}$	Pyroxene	XYZ_2O_6, where X = Mg^{+2}, Fe^{+2}, Mn^{+2}, Li^{+}, Ca^{+2}, Na^{+}, Y = Al^{+3}, Fe^{+3}, Cr^{+3}, Cr^{+3}, Ti^{+4}, Mg^{+2}, Fe^{+2}, Mn^{+2} Z = Si^{+4}, Al^{+3}, Fe^{+3}
Inosilicates double chain $[Si_{4n}O_{11n}]^{6n-}$	Amphibole	$A_{0-1}B_2C_5T_8O_{22}(OH,F,Cl)_2$ A = Na^{+}, K^{+}; B = Na^{+}, Zn^{+2}, Li^{+}, Ca^{+2}, Mn^{+2}, Fe^{2+}, Mg^{+2}; C = Mg^{+2}, Fe^{2+}, Mn^{+2}, Al^{+3}, Fe^{3+}, Ti^{+4}, Zn^{+2}, Cr^{+3}; T = Si^{+4}, Al^{+3}, Ti^{+4}
Phyllosilicates $[Si_{2n}O_{5n}]^{2n-}$	Micas	$X_2Y_{4-6}Z_8O_{20}(OH, F)_4$, X is K, Na, or Ca or less commonly Ba, Rb, or Cs;Y is Al, Mg, or Fe or less commonly Mn, Cr, Ti, Li, etc.;Z is chiefly Si or Al, but also may include Fe^{3+} or Ti.
Tectosilicates $[Al_xSi_yO_{(2x+2y)}]^{x-}$	Feldspar	$KAlSi_3O_8$-$NaAlSi_3O_8$-$CaAl_2Si_2O_8$
	Quartz	SiO_2

Table 2.3 Major secondary minerals and their ideal chemical formula

Type of minerals	*Mineral mame*	*Chemical formula*
Oxides of Fe and Al	Goethite	FeOOH
	Hematite	Fe_2O_3
	Gibbsite	$Al(OH)_3$
Aluminosilicate clays	Kaolinite	$Al_2(OH)_4Si_2O_5$
	Halloysite	$Al_2(OH)_4Si_2O_5.2H_2O$
	Illite (group)	$(K,H)Al_2(Si,Al)_4O_{10}(OH)_2$-$xH_2O$ x = variable amount of water
	Chlorite (group)	$(Mg,Fe)_3(Si,Al)_4O_{10}(OH)_2{\cdot}(Mg,Fe)_3\ (OH)_6$
	Smectite (group)	(Ca,Na,H) $(Al,Mg,Fe,Zn)_2(Si,Al)_4O_{10}(OH)_2$-$xH_2O$ x = variable amount of water
Carbonate and sulfate minerals	Dolomite	$CaMg(CO_3)_2$
	Calcite	$CaCO_3$
	Gypsum	$CaSO_4.2H_2O$

vided into its tri- and dioctahedral minerals according to cation substitutions in the octahedral sheet and within the interlayer (Table 2.4). For example, mica group which has a very similar structure to that of talc and pyrophyllite, except that substitution of Al^{+3} for Si^{+4} in every fourth tetrahedral site. It results in an

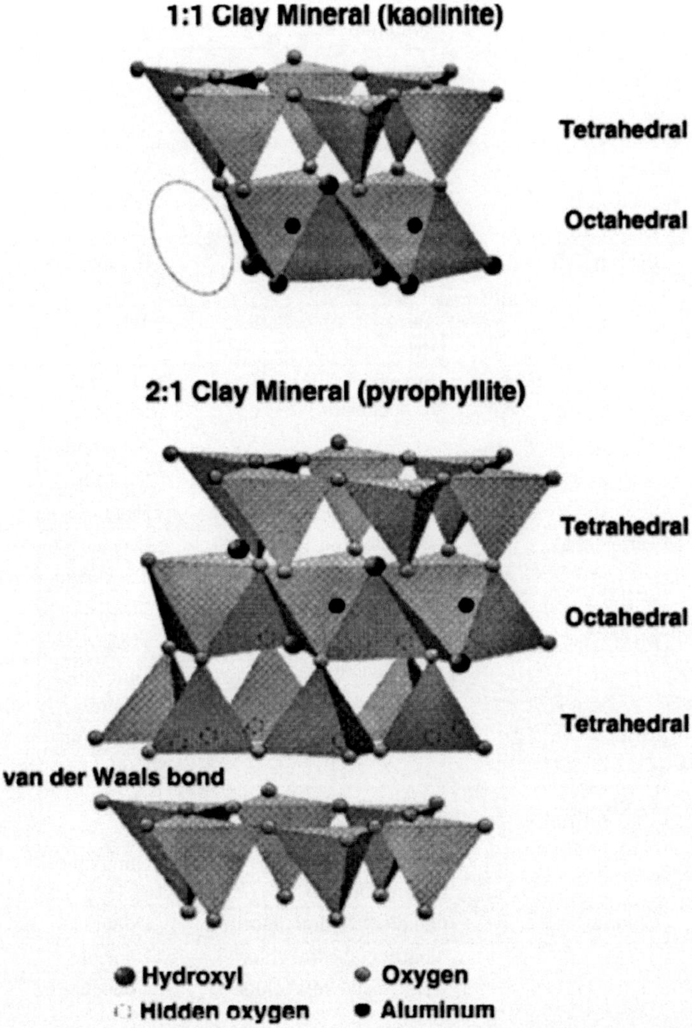

Fig. 2.3 Diagrammatic sketch of clay minerals consisting of: 1:1 clay mineral consisting of one tetrahedral sheet bonded to an octahedral sheet (kaolinite); and 2:1 an octahedral sheet bound between two tetrahedral sheets (pyrophyllite).

excess of one negative charge per formula unit which is satisfied by monovalent cations, primarily K^+, that reside on interlayer sites between the 2:1 layers. Other transition groups are smectite–saponite, vermiculite and Illite.

2:1:1 Clay Minerals In this basic 2:1 structure is intact with octahedral sheet in the interlayer. Chlorites are often referred to as 2:1:1 clays since they are 2:1 clays with a hydroxide interlayer which are octahedrally coordinated. Chlorites can be trioctahedral in both octahedral sheets, i.e., the octahedral sheet of the 2:1 layer and the interlayer octahedral hydroxide sheet, and are referred to as tri,trioctahedral

Table 2.4 Classification of phyllosilicate clay minerals

Type	*Group*	*Subgroup*	*Species*
1:1	Kaoline-serpentine	Kaolins (dioctahedral)	Kaolinite
			Dickite
			Nacrite
			Halloysite
		Sepentines (trioctahedral)	Antigorite
			Chrysotile
2:1	Pyrophyllite-talc	Pyrophyllite (dioctahedral)	Pyrophyllite
		Talc (trioctahedral)	Talc
	Smectite-saponite	Smectites (dioctahedral)	Montmorillonite
			Beidellite
			Nontronite
		Saponite (trioctahedral)	Saponite
			Hectorite
	Vermiculite	Dioctahedral Vermiculite	Dioctahedral Vermiculite
		Trioctahedral Vermiculite	Trioctahedral Vermiculite
	Illite	Illite	Illite
	Mica	Mica (dioctahedral)	Muscovite
			Paragonite
		Mica (trioctahedral)	Biotite
			Phlogopite
			Lepidolite
	Brittle mica	Brittle mica (dioctahedral)	Margarite
		Brittle mica (trioctahedral)	Clintonite
2:1:1	Chlorite	Di,dioctahedral chlorite	Donbassite
		Di,trioctahedral chlorite	Cookeite
		Tri,trioctahedral chlorite	Clinochlore

chlorites. Chlorites that are dioctahedral in the 2:1 layer and trioctahedral in the interlayer hydroxide sheet are referred to as di,trioctahedral chlorites (Spark, 2003).

Other Phyllosilicate Clay Minerals

Hydroxy-Al interlayered vermiculites or smectites are considered to be intermediate mineral weathering products, or from deposition of hydroxy-Al polymeric components within the interlayer space of expanding minerals. Thus, their composition depends totally on their basic 2:1 clay minerals and the interlayer material.

Interstratified clay minerals (or mixed minerals) comprises complex assemblages of primary minerals, and weathering intermediates of multiple structural and compositional combinations. Examples include mica–vermiculite, mica–smectite, chlorite– vermiculite, kaolinite–smectite etc.

PROCESS TO ESTIMATE MINERAL CONTENT

1. Identification of any mineral is done using XRD, by detecting its crystal structure. For example calcite and aragonite has same chemical composition ($CaCO_3$) but different crystal structure.
2. From the basic formula, find out the residue after reaction. Case studies: a Calcite and b. Gypsum.

For Calcite

$CaCO_3 = CaO + CO_2$
100.0956.08 44.01
Percentage of CaO in $CaCO_3$ = (56.0774/100.0869) × 100 ≈ 56.029%
Percentage of Ca in CaO = (40.078/56.0774) × 100 ≈ 71.469%

For Gypsum

Gypsum on heating at 373K forms Plaster of Paris
$CaSO_4.2H_2O = CaSO_4.1/2\ H_2O + 1\ 1/2\ H_2O$
172.17 145.15 27.02
The weight loss for the conversion of gypsum to plaster of paris = 1/2 H_2O = 9.00764
If 172.17 g of gypsum is heated to form plaster of paris then residue weight = 172.1 7 – 9.00764 = 163.16
Percentage wt. loss at 373K = (9./172.17) × 100 = 5.22% when sample is converted to plaster of paris.
After 10 g of gypsum is heated at 373 K to form plaster of paris, weight loss = 0.3 g

It means, whole sample has not been converted. This maybe due to sample preparation (it must be homogenous), heating rate for the analysis is too fast. These are the precautions needed for the experiment.

1. Using thermal analysis, the decomposition residue can be estimated. Decomposition temperature depends on the chemical composition as well as crystal structure of the sample.

For example, calcite and aragonite having same chemical composition but different crystal structure, has large difference in decomposition temperature, i.e calcite (≈884 °C) and aragonite (≈280 °C).

CONCLUSION

Study of primary and secondary minerals are extremely important as they not only provide information about soil development but also serve as sources of plant nutrients. Since clay minerals constitutes <2 μm size of particles which mainly include sand and silt, their structural study have been performed in detail. The secondary clay minerals (phyllosilicates) in soils play a profound role in affecting numerous soil chemical reactions and processes which have been discussed in following chapters.

THINK FOR A WHILE

- Why secondary minerals have smaller particle size?
- What is the role of clay minerals in soil science?
- What is the difference between dioctahedra and trioctahedra sheets in clay minerals?
- What is the difference between 2:1 and 2:1:1 clay minerals?
- What kind of clay minerals are vermaculite and smectites?
- What will be the residue left after heating 10 gm of dolomite?

REFERENCES

Barton, C.D. and Karathanasis, A.D. (2002). Clay Minerals. *In:* Lal, R., Ed., Encyclopedia of Soil Science, Marcel Dekker, New York, USA, 187–192.

Spark, D.L. (2003). Environmental Soil Chemistry. Academic Press.

Chapter 3
Soil Taxonomy and Classification

Soil taxonomy is a basic system by which soil is classified into various classes. Its main objective is to understand the relationship among and between soils and the factors responsible for their character. It was mainly used for soil survey but modifications with time also made it an important means of communication for the discipline of soil science in agriculture, forestry and engineering to predict soil behaviour under defined use and management or manipulation.

Soil classification includes taxonomy and grouping of soil according to their purpose. Different purposes have different classification and thus would have no drawbacks.

Soil taxonomy is classified by different properties of soil such as soil depth, moisture, temperature, texture, structure, cation exchange capacity, base saturation, clay mineralogy, organic matter content and salt content, out of which, soil horizons are the primary building blocks.

Diagnostic surface horizons or epipedons are formed at the soil surface whereas subsurface soil horizons are formed below the soil surface. There are currently eight diagnostic epipedons (Table 3.1) and the 20 diagnostic subsurface horizons (Table 3.2) in Soil Taxonomy. They are defined mostly in terms of soil colour, content of organic matter and base saturation; i.e., relative amounts of bases such as calcium, magnesium, sodium and potassium as compared with hydrogen.

CATEGORIES

Among other classification systems, the USDA soil taxonomy system developed by United States Department of Agriculture and the National Cooperative Soil Survey in 1975 is most widely used. According to the classification, *soil taxonomy* has six categories, namely, order, suborder, great group, subgroup, family, and series (Fig. 3.1.).

S. Mukherjee, *Current Topics in Soil Science*,
https://doi.org/10.1007/978-3-030-92669-4_3

Table 3.1 Eight diagnostic surface horizons/epipedons

Epipedons	*Description*
Anthropic (Gr. *anthropos*, human being)	Manmade, thick, dark, formed by long continued residence or application of organic residues. Often with high phosphorus content
Folistic (L. *folia*, leaf)	More than 20 cm thick, high in organic matter of cool humid regions under forest vegetation
Histic (Gr. *histos*, tissue)	Between 20 and 40 cm thick, high in organic matter that is saturated with water for long periods
Melanic (Gr. *melas*, black)	Thick, black, high in organic matter of soils formed from volcanic ash. The organic matter is associated with short-range order minerals or Al–humus complexes
Mollic (L. *mollis*, soft)	Thick, dark, high in organic matter, well structured with over 50% base saturation, predominantly from divalent cations. Typical for steppe soils
Ochric (Gr. *ochros*, pale)	Thin, light, or low in organic matter, thereby failing to be mollic or umbric
Plaggen (Ger. *plaggen*, sod)	Manmade, more than 50 cm thick, high in organic matter, and raised above the original soil surface. Formed by spreading of a mixture of manure and sod, the latter used for bedding livestock in medieval times
Umbric (L. *umbra*, shade/dark)	Thick, dark, high in organic matter, well structured with under 50% base saturation

Gr: Greek, L: Latin, Ger: German
Based on Krogh (2006)

ORDER

There are 12 soil orders (Table 3.3) and they are categorized based on physical, chemical and biological properties. These properties reflect greatly on the genesis of soil of different orders. They are separated on the basis of stable and long-term development processes and properties reflected by diagnostic horizons.

SUBORDER

Within the order, there are 64 suborders recognized which are separated on the basis of mostly dynamic properties controlling the current soil-forming processes such as soil moisture regime and temperature regime.

The soil moisture regime (Table 3.4) is an expression of the changes in soil water over time. It refers to the presence or absence of either groundwater or of water held at a potential of less or more than −1,500 kPa (–15 bars) by periods of the year. Three soil water conditions are recognized: soil that is water saturated (wet soil), soil between saturation and the permanent wilting point (moist soil), and soil drier than the wilting point (dry soil) (Foth, 1990).

Table 3.2 Twenty diagnostic subsurface horizons

Subsurface horizon	*Description*
Agric (L. *ager*, field)	Illuvial horizon below the plow layer containing silt, clay, and organic matter. Formed under (old-world) cultivation.
Albic (L. *albus*, white)	Eluvial, bleached and light-coloured horizon from which clay and Fe-oxides have been removed.
Argillic (L. *argilla*, white clay)	Horizon with significantly higher clay content than horizon above. Formed through clay illuviation and/or in situ weathering.
Calcic (L. *calcis*, lime)	Illuvial horizon with secondary $CaCO_3$ as powder or concretions. Mostly found in arid environments.
Cambic (L. *cambiare*, to exchange)	Horizon in which weak structure formation, weathering of primary minerals, and precipitation of Fe-oxides have taken place.
Duripan (L. *durus*, hard)	Horizon cemented by illuvial silica thereby impeding root and water movement.
Fragipan (L. *fragilis*, brittle)	Compact, high bulk density, noncemented horizon impeding root and water movement.
Glossic (Gr. *glossa*, tongue)	Transitional horizon with vertical bleached tongues formed by movement and destruction of clay and Fe-oxides in existing argillic horizon.
Gypsic (L. *gypsum*, gypsum)	Illuvial horizon with secondary gypsum. Mostly found in arid environments.
Kandic (from mineral kandite)	Horizon with significantly higher clay content than horizon above and dominated by LAC clays. Formed through clay illuviation and/or in situ weathering.
Natric (L. *natrium*, sodium)	Horizon with significantly higher clay content than horizon above and 15% or more exchangeable Na. Formed through clay illuviation and/or in situ weathering.
Ortstein (Ger.)	Cemented horizon consisting of complexes of Al and organic matter with or without Fe (spodic materials).
Oxic (F. oxide)	Fine textured, thoroughly weathered horizon with clay consisting of LAC and sesquioxides and having a low CEC. Mostly found in old tropical environments.
Petrocalcic (Gr. *Petra*, rock and calcic)	Indurated illuvial horizon with secondary $CaCO_3$. Mostly found in arid environments. Barrier to roots and water.
Petrogypsic (Gr. *petra*, rock and gypsic)	Indurated illuvial horizon with secondary gypsum. Mostly found in arid environments. Barrier to roots and water.
Salic (L. *sal*, salt)	Horizon with accumulation of salts more soluble than gypsum and in quantities harmful to roots. Mostly found in arid environments.
Sombric (F. *sombre*, dark)	Horizon with illuvial concentrations of organic matter that is not in combination with Al. Mostly found in cool, mountainous environments.
Spodic (Gr. *spodos*, wood ash)	Illuvial horizon with amorphous material consisting of organic matter and Al with or without Fe.
Sulfuric (L. *sulphur*, sulfur)	More than 15 cm thick mineral or organic horizon with low pH (pH 3.5) formed by artificial drainage of sulfide rich materials.
Placic (Gr. *plax*, flat stone)	Thin (<25 mm), black to reddish horizon cemented by Fe (and Mn) and organic matter.

Gr: Greek, L: Latin, Ger: German
Based on Krogh, 2006

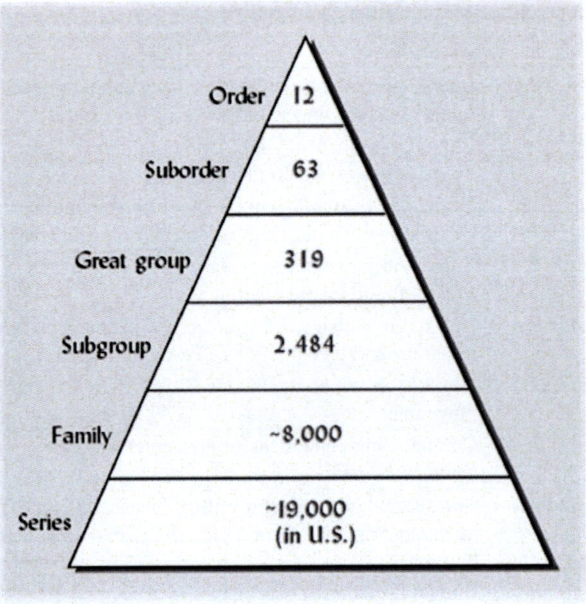

Fig. 3.1 Categories of soil taxonomy.

Soil temperature regimes (Table 3.5) are defined according to the mean annual soil temperature, mean summer temperature, and the difference between mean summer and winter temperatures in the root zone (arbitrarily set at 5 to 100 cm).

The suborder name has two syllables. First connotes something of the diagnostic properties of the soil, and the second is the formative element or syllable for the order. For example xeralfs means alfisols with xeric moisture regime or ochrepts means inceptisols with ochric epipedon.

GREAT GROUP

Suborders are again divided into great groups. There are about 300 great groups separated on the basis of generally static properties with minor or additional control on current soil-forming processes. Horizons marked by clay, iron, humus and hard pans and soil features such as the expansion-contraction of clays (that produce self-mixing provided by clay), temperature, and marked quantities of various salts are used as distinguishing features (Donahue et al., 1977).

Great group names are formed simply by putting a great group formative element in front of the suborder name. The prefix is connotative and suggests the unique features of the particular great group named (Table 3.6). For example, durixeralfs means suborder xeralfs with duripan.

Table 3.3 Twenty soil orders and their characteristics

Name	*Major characteristics*	*Name*	*Major characteristics*
Alfisols	Must have argillic, natric, or kandic horizon High-to-medium base saturation Moderately weathered Commonly form under boreal or broadleaf forests Rich in iron and aluminum Common in humid areas, semi-tropics, and mediterranean climates 9.6% of global and 14.5% of U.S. ice-free land	Andisols	Form from volcanic ejecta, dominated by allophane or Al-humic complexes Must have andic soil properties: high in poorly crystalline Fe and Al minerals, high in phosphorus, low bulk density, and high proportions of glass and amorphous colloidal materials, such as allophane, imogolite and ferrihydrite High Organic Matter content, sometimes melanic epipedon 0.7% of global and 1.7% of U.S. ice-free land
Aridisols	Dry soil (i.e., must have aridic moisture regime). Ochric epipedon is common; Sometimes argillic or natric horizon; Must have some diagnostic subsurface horizon Commonly in deserts 12.7% of global and 8.8% of U.S. ice-free land	Entisols	Least soil profile development; ochric epipedon is common; no B horizons Most common order by surface area (16.3% of global and 12.2% of U.S. ice-free land)
Gelisols	Soils with permafrost within 100 cm or cryoturbation (frost churning) within 100 cm plus permafrost within 200 cm Commonly at high latitudes and elevations 8.6% of global and 7.5% of U.S. ice-free land	Histisols	Must have histic epipedon Usually aquic soil moisture regime No diagnostic subsurface horizons Rapid decomposition when aerated Peat or bog; >20% organic matter; organic soil materials extending down to an impermeable layer or with an organic layer that is more than 40 cm thick and without andic properties Commonly in wetlands (swamps, marshes, etc.) 1.2% of global and 1.3% of U.S. ice-free land

(continued)

Table 3.3 (continued)

Name	*Major characteristics*	*Name*	*Major characteristics*
Inceptisols	Similar to Entisol, but beginning of a B horizon is evident; no diagnostic subsurface horizons On landscapes continuously eroded or young deposits Cambic, sulfuric, calcic, gypsic, petrocalcic, or petrogypsic horizon, or with a mollic, umbric, or histic epipedon, or with an exchangeable sodium percentage of >15% or fragipan 9.9% of global and 9.1% of U.S. ice-free land	Mollisols	Must have mollic epipedon High base saturation of >50% Dark soils; Some with argillic or natric horizons Common in grasslands 6.9% of global and 22.4% of U.S. ice-free land
Oxisols	Most soil profile development; Must have oxic horizon within 150 cm of soil surface Low nutrient availability No argillic horizon Highly weathered Dominated by end-member clays, Al and Fe oxides Commonly in old landscapes in tropics 7.6% of global and <0.01% of U.S. ice-free land	Spodosols	Must have spodic horizon within 2 m of soil surface and without andic properties Usually have albic horizon High in Fe, Al oxides and humus accumulation; Acidic soils Common in coniferous or boreal forests 2.6% of global and 3.3% of U.S. ice-free land
Ultisols	Must have argillic or kandic horizon Low base saturation of <35% at 2 m depth or 75 cm below a fragipan Common in subtropical regions; often known as red clay soils 8.5% of global and 9.6% of U.S. ice-free land	Vertisols	Usually mollic epipedon High in shrinking and swelling clays; >30% clay to a depth of 50 cm; deep cracks form when soil dries; form from parent material high in clay (e.g., shales, basins, exposed Bt horizons of old soils) 2.4% of global and 1.7% of U.S. ice-free land

Based on USDA Classification of Soil Taxonomy

SUBGROUP

Great groups are further subdivided into subgroups. There are about 2400 subgroups separated on the basis of a central concept and divergence. Three types of subgroups are:

- **Typic** Subgroup that represents the soils that do not have the characteristics defined for the other subgroups i.e. typical in nature.
- **Intergrades** These describes the properties that suggest how it grades towards soils of other soil great groups, suborders or orders. These properties may be the result of processes that cause one kind of soil to develop from or toward another

Table 3.4 Soil moisture regimes

Soil moisture regime	*Major characteristics*
Aquic	Soil is saturated with water and virtually free of gaseous oxygen for sufficient periods of time, such that there is evidence of poor aeration Common in wetlands
Udic	Soil moisture is sufficiently high year-round in most years to meet plant requirement Common in humid regions
Ustic	Soil moisture is intermediate between Udic and Aridic regimes Plant-available moisture during the growing season, but severe periods of drought may occur Common in semi-arid regions
Aridic	Soil is dry for at least half of the growing season and moist for less than 90 consecutive days Common in arid (desert-like) regions
Xeric	Soil moisture regime is found in Mediterranean-type climates, with cool, moist winters and warm, dry summers Like the Ustic Regime, it is characterized as having long periods of drought in the summer

Table 3.5 Soil temperature regimes

Soil temperature regime	*Temperature range*
Pergelic	~ –8 °C to –4 °C
Subgelic	~ –4 °C to 0 °C
Frigid	~ 0 °C to 8 °C
Mesic	8 °C to 15 °C
Thermic	15 °C to 22 °C
Hyperthermic	22 °C or higher

kind of soil or otherwise to have intermediate properties between those of two or three great groups.

- **Extragrades** These have some properties that are not representative of the great group but that do not indicate transitions to any other known kind of soil, e.g. soils that have high amounts of organic carbon (Humic) (USDA, 1999).

Names of subgroups are binomial. Each binomial has an adjective before the name of the great group to which the subgroup belongs, e.g. typic durixeralfs.

Table 3.6 Prefixes and their connotation for great groups

Prefix	*Connotation*	*Prefix*	*Connotation*
acr	Extreme weathering	luv	Illuvial
agr	An agric horizon	med	Of temperate climates
alb	An albic horizon	nadur	Combination of natr and dur
and	Ando-like	natr	Presence of natric horizon
anthr	An anthropic epipedon	ochr	Presence of ochric epipedon
arg	An argillic horizon	pale	Old development
bor	Cool	pell	Low chroma
calc	An calcic horizon	plac	Presence of thin cemented layer
camb	An cambic horizon	plag	Presence of plaggen horizon
chrom	High chroma	plinth	Presence of plinthite
cry	Cold	psamm	Sand textures
dur	A duripan	quartz	High quartz content
dystr/dys	Low basic cation saturation	rhod	Dark red color
eutr/eu	High basic cation saturation	sal	Presence of salic horizon
ferr	Presence of iron	sider	Presence of free iron oxides
fluv	Floodplain	sombr	A dark horizon
frag	Presence of fragipan	sphagn	Presence of sphagnum moss
fragloss	Combination of frag and gloss	sulf	Presence of sulfides or their oxidation products
gibbs	Presence of gibbsite	torr	Aridic moisture regime
gloss	Tongued	trop	Continually warm and humid
gyps	Presence of gypsic horizon	ud	Udic moisture regime
hal	Salty	umbr	Presence of umbric epipedon
hapl	Minimum horizon	ust	Ustic moisture regime
hum	Presence of humus	verm	Wormy or mixed by animals
hydr	Presence of water	vitr	Presence of glass
kand	Presence of low activity clay	xer	Xeric moisture regime

Foth, 1990

FAMILY

A soil family category is a group of soils within a subgroup and describes the physical and chemical properties which affect the response of soil to agricultural management and engineering applications. The principal characteristics used to differentiate soil families include texture, mineralogy, pH, permeability, structure, consistency, local precipitation pattern, and soil temperature. For some soils the criteria also specify the percentage of silt, sand and coarse fragments such as gravel, cobbles and rocks (Donahue et al., 1977). There are about 8000 soil families known currently.

Family names are fairly common terms that denote in more detail the features of a particular subgroup, e.g. a fine, montmorillonitic, thermic soil family is one that is fine textured, contains montmorillonite as the dominant mineral in the clay and has a warm, or thermic mean annual soil temperature.

SERIES

Series are the lowest category of soil taxonomy. There are about 20000 series known. They are the mapping units carrying local names and are separated on the basis of properties that reflect relatively narrow ranges of soil-forming factors. Examples include the Gila, Graham, Mohave, Tubac and Moenkopie series. Other names are variants of these place names and still others simply are coined.

Order: Alfisols
Suborder: Xeralfs
Great Group: Durixeralfs
Subgroup: Abruptic Durixeralfs
Family: Fine, Mixed, Active, thermic Abruptic Durixeralfs
Series: San Joaquin

CONCLUSION

The major difficulty in classifying soils is that unlike plants and animals, soils are a continuum in which properties may change gradually with distance. In an attempt to solve this problem soil taxonomy was constructed that describes even the smallest volume of soil (pedon). But with increasing demands on small scale quantitative soil information and a reality where soil information must be able to address a broad range of environmental issues other than just agricultural production. The soil research community is faced with a major challenge in meeting this demand and think along new lines to develop new soil information systems.

THINK FOR A WHILE

- What is the need for soil taxonomy? Why is it difficult to classify soils?
- How did the six categories under soil taxonomy come into existence?
- What will be the order and properties of following suborders?

(a) Fluvents
(b) Humods
(c) Vitrands

- Identify the soil categories and properties from soil subgroups?
  (a) Andic Humixerept
  (b) Aquic Durudand
  (c) Xeric Palehumult

REFERENCES

Donahue, R.L., Miller, R.W. and Shickluna, J.C. (1977). Soils: An Introduction to Soils and Plant Growth. Prentice-Hall. Upper Saddle River.

Foth, H.D. (1990). Fundamentals of Soil Science. John Wiley & Sons.

Krogh, L. (2006). Classification Systems : Major. Taylor & Francis.

United States Department of Agriculture. Soil Taxonomy: A Basic System of Soil Classification for Making and Interpreting Soil Surveys. (1999). By Soil Survey Staff. Agriculture Handbook Number 436.

Chapter 4
Soil Quality

Soil quality is a measure of the capacity of a soil to perform necessary functions. Definitions of soil quality depends on human uses of soil and functions of soil within natural and agricultural ecosystem. Functions of soil includes maintaining biodiversity and productivity, partitioning water and solute flow, filtering and buffering, nutrient cycling, and providing support for plants, human health and habitation.

IMPORTANCE

The concept of soil quality is still continuing to evolve. It is different for every field. In agriculture, soil quality means productive land that maintains or increases farm profitability, as well as the conservation of farm resources for future farming. For a consumer, it mainly concerns with dependable supply of wholesome, affordable food and the protection of drinking water quality. To the environmentalist, it may be defined in terms of soil functioning at its potential in an ecosystem with respect to maintenance or enhancement of biodiversity, water quality, nutrient cycling, and biomass production.

Recently soil quality has been given an immense attention. This is because of the concern that soil quality may be declining. Since soil itself is finite and nonrenewable, sustainability of human uses of soil is a global concern. Soil erosion, protecting natural areas, soil contamination and maintaining agriculture productivity are few of the many issues being faced. Thus, measurement of soil quality is needed. With the data acquired from such measurements, restoration, conservation and management practices of soil will be responded. Since there are many functions of soil, as discussed before, soil quality is also not measured by any single conserving or degrading process or property.

S. Mukherjee, *Current Topics in Soil Science*,
https://doi.org/10.1007/978-3-030-92669-4_4

SOIL QUALITY AND SOIL HEALTH

Soil quality and soil health are terms often used synonymously but some researchers considers soil health as a subset of soil quality. With stress on that the focus of soil health is on the organic and more dynamic soil properties. Soil health is defined as the continued capacity of soil to function as a vital living system, by recognizing that it contains biological elements that are key to ecosystem function within land-use boundaries (Doran and Zeiss, 2000; Karlen et al., 2001), i.e. soil health mainly concerns with state of the soil at particular time whereas soil quality may be considered as soil usefulness for a particular purpose.

Soil health indicators too are needed like soil quality but they are specially for small holder farmers to understand the chain of cause and effect that links farm decisions to ultimate productivity and health of plants and animals. Soil health is evaluated by comparing the present state of soil with a set of reference values. The reference values comes from three types of soils, viz. the native soil, intensively managed soil and altered or reclaimed soils.

Soil health and quality indicators are of three types (Fig. 4.1, Table 4.1):

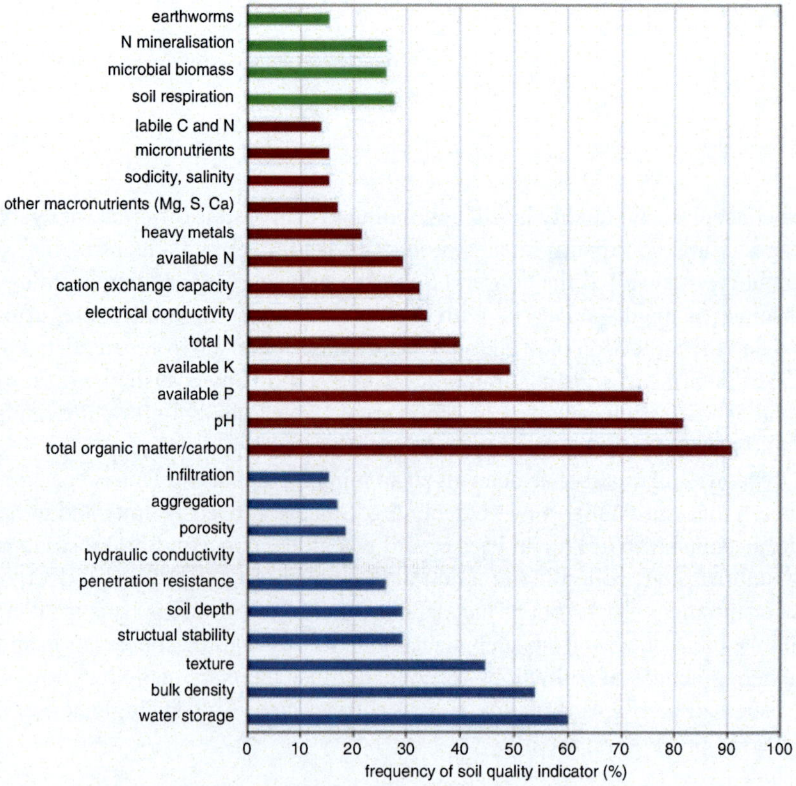

Fig. 4.1 Frequency of different biological, chemical and physical indicators (%) used in soil quality assessment. Soil biological, chemical and physical indicators are shown in green, red and blue, respectively.

Table 4.1 Soil physical, chemical and biological properties used as indicators of soil quality

Soil property	*Method*	*Information*
Physical		
Soil texture	Hydrometer method	Retention and transport of water, nutrients, and chemicals Susceptibility to erosion Stabilization of organic matter and soil structure
Topsoil and rooting depth	Soil coring Excavation	Water and nutrient availability Rooting volume for crop production
Soil bulk density	Soil core Neutron probe	Volume of pore space Compaction
Infiltration and hydraulic conductivity	Pressure, tension infiltrometer Permeameter; laboratory measurement on cores	Runoff and leaching potential Drainage
Water content	Gravimetric measurement by oven drying Volumetric measurement by TDRa or core	Available water
Water holding capacity	Water content at 33 and 1500 kPa tension	Availability of air and water Retention and transport of water and chemicals Drainage
Chemical		
pH	pH meter	Acidity or alkalinity of soil Nutrient availability
Electrical conductivity	Conductivity meter	Presence and quantity of soluble salts
Available N, P and K	Laboratory extraction and analysis	Plant available nutrients
Organic C and N	Wet or dry combustion	Organic matter reserves Nutrient cycling Soil structure
Biological		
Potentially mineralizable N	Aerobic or anaerobic incubation	Potential to supply plant-available N
Microbial biomass	Chloroform fumigation followed by incubation or extraction	Size of microbiological population Pool of rapidly cycling organic matter and nutrients
Soil respiration	In situ and laboratory measurements by CO_2 gas analyzer or alkali traps	Availability of soil organic matter reserves Microbiological activity

Adapted from Doran and Parkin (1994) and Gregorich (2006)

- **Physical Indicators** – Soil's texture, bulk density, porosity, permeability, hydraulic conductivity are mainly considered physical indicators of soil quality. Often they cannot be easily improved. Scientists consider many physical properties like texture, structure and bulk density to not effect plant productivity rather water potential; oxygen diffusion rate and temperature are the best indicators for plant production. Soil depth is also considered one of the best indicators for soil quality as it provides information about depth available for roots to explore for water and nutrients.
- **Chemical Indicators** – Nutrients are needed for high crop yields and for that chemical tests are done to understand the capacity of soil for supplying nutrients. These mainly depend on pH and CEC, e.g. in very acidic soils, liming is required to increase the pH of the soil. Thus, nutrients like Ca and Mg are supplied in terms of inorganic fertilizers to bring the soil back to neutral. Soil properties can be compromised by human interference that can increase the toxicity of soil. Thus, contaminant studies are done to understand the risk and finding what measures should be taken.
- **Biological Indicators** – Soil quality is strongly influenced by microbiological mediated processes like nutrient cycling, nutrient capacity, aggregate stability etc. Thus, indicators like soil organic matter, respiration, microbial biomass, mineralizable nitrogen and phosphorus are measured that influences microbiota and soil fauna directly. High rates of nitrogen turnover indicates healthy soil biological community whereas soil organic matter determines water holding capacity and susceptibility of soil to degradation. It may also serve as a source or sink to atmospheric CO_2. Increase in the soil carbon content is indicated by a higher microbial biomass and elevated respiration.

MEASUREMENT

Measurement of soil quality is done with the help of Minimum Data Set (MDS). It is a minimum set of indicators required to obtain a comprehensive understanding of the soil attributes required for a specific purpose and is not feasible to measure characteristics of all soil qualities. Thus, indicators need to be selected to address the issues of greatest concern (Fig. 4.2). The main concern in assessing soil quality are: soil erosion and redistribution; chemical and biological contamination; and soil degradation and depletion. The key indicators often used in MDS are soil texture, organic matter, pH, nutrient status, bulk density, electrical conductivity and rooting depth. Selecting an appropriate suite of indicators for a particular soil quality assessment is an important first step in monitoring soil quality. Since the importance of soil quality indicators changes temporally and spatially, MDS also changes with them. For example, for farmers these tools need to be simple measures of soil health and soil quality. For extension and policy personnel, they provide basic information needed to arrive at management decisions (Barrios et al., 2006). For researchers, there is need to conduct sufficiently detailed tests while controlling for variation in order to develop meaningful assessments of soil status, often expressed as an index of soil quality (Kang et al., 2005).

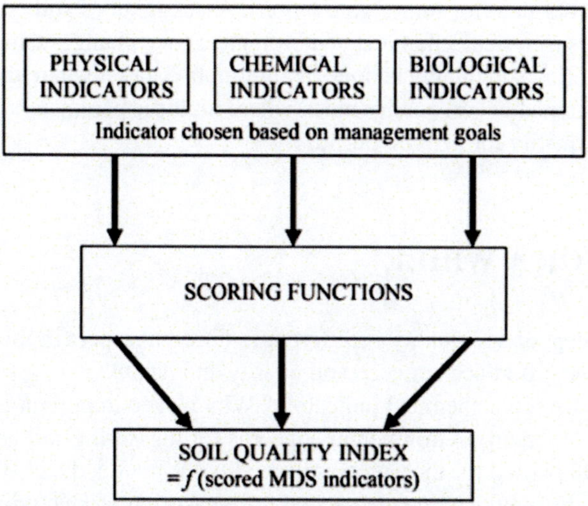

Fig. 4.2 Flow chart indicating assessment of soil quality through MDS.

SOIL QUALITY INDEX

Common approaches utilized for the evaluation of soil quality are the use of different individual soil parameters or single indices (e.g. metabolic quotient, the metabolic ratio). However, the use of one individual indicator or indices integrating only two parameters has many limitations and provides insufficient information about soil quality and degradation (Bastida et al., 2008; Masto et al., 2015). Therefore, derivation of complex, multiparametric indices combining different parameters were needed. A soil quality index (SQI) could be defined as a minimum set of parameters that provides numerical data concerning the capacity of a soil to carry out one or more functions (Garrigues et al., 2012; Asensio et al., 2013) and is represented by the equation:

$$SQI = f\left(SP, P, E, H, ER, BD, FQ, MI\right)$$

Where *SP* indicates soil properties, *P* potential productivity, *E* environmental factors, *H* human and animal health, *ER* erodibility, *BD* biological diversity, *FQ* food quality and safety and *MI* management inputs.

CONCLUSION

A soil management system is considered sustainable only if it maintains or improves soil quality and does not compromise on environmental quality. For that understanding, soil quality is important to achieve environmental protection and preserve soils in a usable state for future generations. Assessment of soil quality is done by soil indicators (physical, chemical and biological) that changes temporally and spatially. Monitoring soil quality is done by minimum data set where individual parameters are combined or soil quality index where multiple parameters are combined to measure the quality for definite purpose.

THINK FOR A WHILE

- Why concept of soil quality is different in different aspect in life?
- What is the difference between soil quality and health?
- What are the main chemical indicators? Why is it so important in agriculture?
- What are the main environmental concerns for measuring soil quality?
- Why MDS is used for measuring soil quality? Why it is different from SQI?
- What are the main parameters needed for measuring productivity in a farm?

REFERENCES

Asensio, V., Guala, S.D., Vega, F.L. and Covelo, E.F. (2013). A soil quality index for reclaimed mine soils. *Environ. Toxicol. Chem.,* **32:** 2240–2248.

Barrios, E., Delve, R.J., Bekunda, M., Mowo, J., Agunda, J., Ramisch, J., Trejo, M.T. and Thomas, R.J. (2006). Indicators of soil quality: A South-South development of a methodological guide for linking local and technical knowledge. *Geoderma,* **135:** 248–259.

Bastida, F., Zsolnay, A., Hernández, T. and García, C. (2008). Past, present and future of soil quality indices: A biological perspective. *Geoderma,* **147:** 159–171.

Doran, J.W. and Parkin, T.B. (1994). Defining and Assessing Soil Quality. *In:* Doran, J.W., Coleman, D.C., Bezdicek, D.F. and Stewart, B.A., Eds. Defining Soil Quality for a Sustainable Environment, *Soil Science Society of America Journal*, Madison, 3–21.

Doran, John W. and Zeiss, Michael R. (2000). "Soil health and sustainability: Managing the biotic component of soil quality". Agronomy & Horticulture. Faculty Publications. 15.

Garrigues, E., Corson, M.S., Angers, D.A., van der Werf, H.M.G. and Walter, Ch. (2012). Soil quality in Life Cycle Assessment: Towards development of an indicator. *Ecol. Indicators,* **18:** 434–442.

Gregorich, E.G. (2006). Quality. Taylor & Francis.

Kang, G.S., Beri, V., Sidhu, B.S., Rupela, O.P. (2005). A new index to assess soil quality and sustainability of wheat-based cropping systems. *Biology and Fertility of Soils*, **41:** 389–398.

Karlen, Douglas, Andrews, S.S. and Doran, J.W. (2001). Soil quality: Current concepts and applications. *Advances in Agronomy.* **74:** 1–40.

Masto, R.E., Sheik, S., Nehru, G., Selvi, V.A., George, J. and Ram, L.C. (2015). Assessment of environmental soil quality around Sonepour Bazari mine of Raniganj coalfield, India. *Solid Earth,* **6:** 811–821.

Chapter 5
Soil Analytical Techniques

Data obtained from different analytical techniques yields different aspects of the sample property. It can be correlated to get the needed information leading to the knowledge about the sample status.

Information is the data in context or combined with other data, sufficient to support decisions and conclusions.

Knowledge is the interpretation of information. Interpretation produces knowledge which is useful for making decisions. These decisions lead to actions that affect the process.

Analytical techniques to be used depends on the type of problems. It is a specified time bound routine type which can be done online. It needs pre-calibration of the specific analysis, and yields time bound output data (Fig. 5.1).

Important Features of Analytical Techniques:

1. General use
2. Common applications
3. Samples: (i) state, (ii) amount and (iii) preparation. Procedure: In laboratory analysis, samples may be treated, or prepared in a specific way and then used to get analytical measurements.
4. Analysis time
5. Limitations
6. Complimentary or related techniques

To understand different analytical techniques, information needed are:

(A) Brief history
(B) Current use
(C) How it works: Physical and chemical principles, instrumentation (modular approach), Description of each major component, Operation procedure
(D) What it does: Analytical information: (i) Qualitative, (ii) Quantitative, (iii) Accuracy and precision and (iv) Detection limits.
(E) Applications: (i) General, (ii) specific examples

S. Mukherjee, *Current Topics in Soil Science*,
https://doi.org/10.1007/978-3-030-92669-4_5

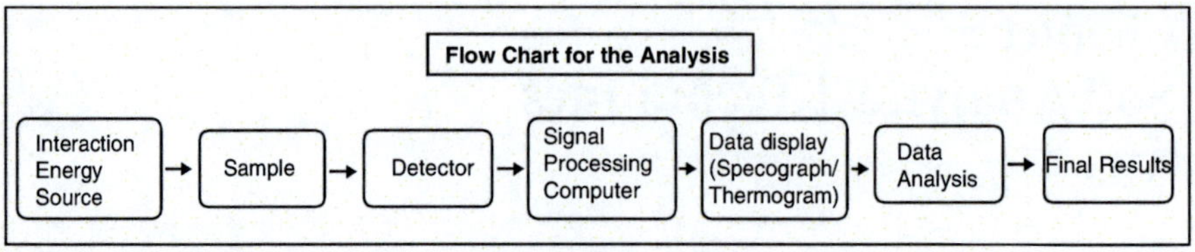

Fig. 5.1 General flow chart for any analytical techniques.

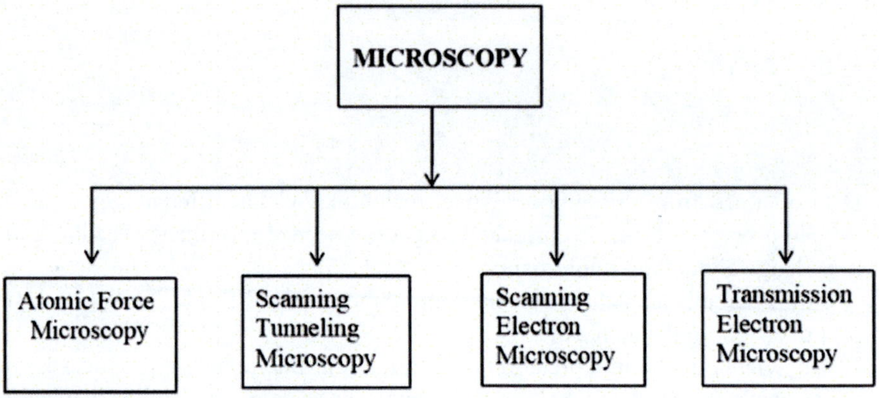

Fig. 5.2 Different microscopic techniques.

Samples and Specimen

- Sample is the material supplied for analysis
- Specimen is the portion of the sample that is prepared and presented to the instrument

How a specimen will be prepared is determined whether it is representative of the sample as a whole and if the resultant data is similarly representative.

There are two types of techniques used, spectroscopic and microscopic.

Microscopic Analytical Techniques
Four most useful microscopic techniques used to characterize materials are (Fig. 5.2):

Electron Probe Micro-Analyser (EPMA) is also a type of microscopic analytical technique.

MICROSCOPIC ANALYSIS

Microscopy is the technical field of using microscopes to view samples and objects that cannot be seen with the unaided eye. The basic function of a electron microscope is as follows (Fig. 5.3):

ELECTRON MICROSCOPY

Electron microscopy (EM) is a technique for obtaining high resolution images of specimens by using a beam of accelerated electrons as a source of illumination. Since the wavelength due to an electron can be much shorter than that of visible light photons, electron microscopes have a higher resolving power than light microscopes and can reveal the structure of smaller objects.

Two main types of electron microscope are the transmission electron microscope (TEM) and the scanning electron microscope (SEM). The TEM uses a high voltage electron beam to illuminate the specimen and create an image. It mainly consists of (i) an electron gun, which produces the electron beam, and the condenser system, which focuses the beam onto the object, (ii) the image-producing system, consisting of the objective lens, movable specimen stage, and intermediate and projector lenses, which focus the electrons passing through the specimen to form a real, highly magnified image, and (iii) the image-recording system, which converts the electron image into some form perceptible to the human eye (Fig. 5.4) (Ford, 2002). TEM can image specimens up to 1 micrometre in thickness.

The SEM uses a beam of electrons which scans over the surface of a solid object and builds up an image of the details of the surface structure. This is why SEM only shows the morphology of samples whereas TEM provides the details about internal composition such as morphology, crystallization, stress or even magnetic domains. The electron source and electromagnetic lenses that generate and focus the beam are similar to those of TEM. The action of the electron beam stimulates emission of high-energy back-scattered electrons and low-energy secondary electrons from the surface of the specimen. Unlike TEM, large and bulky specimens can be accommodated in SEM.

Spectroscopic Analysis

Using optical, electronic properties, crystallinity, elemental and chemical composition, different spectral emissions are observed. Various spectral regions have their typical properties which are utilized for spectroscopic analysis (Figs 5.6 and 5.7). Four main spectroscopic techniques are used as given in Fig. 5.5.

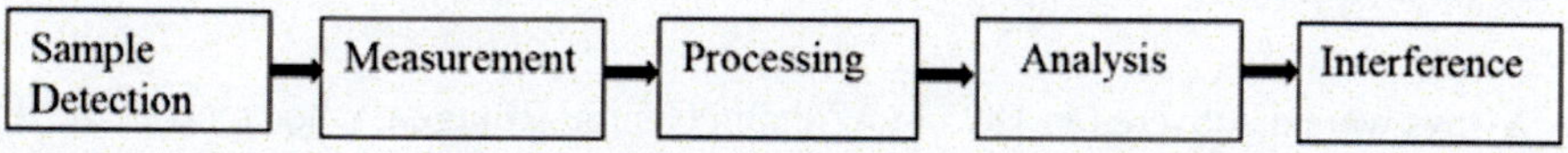

Fig. 5.3 Flow chart to understand the data for final inference.

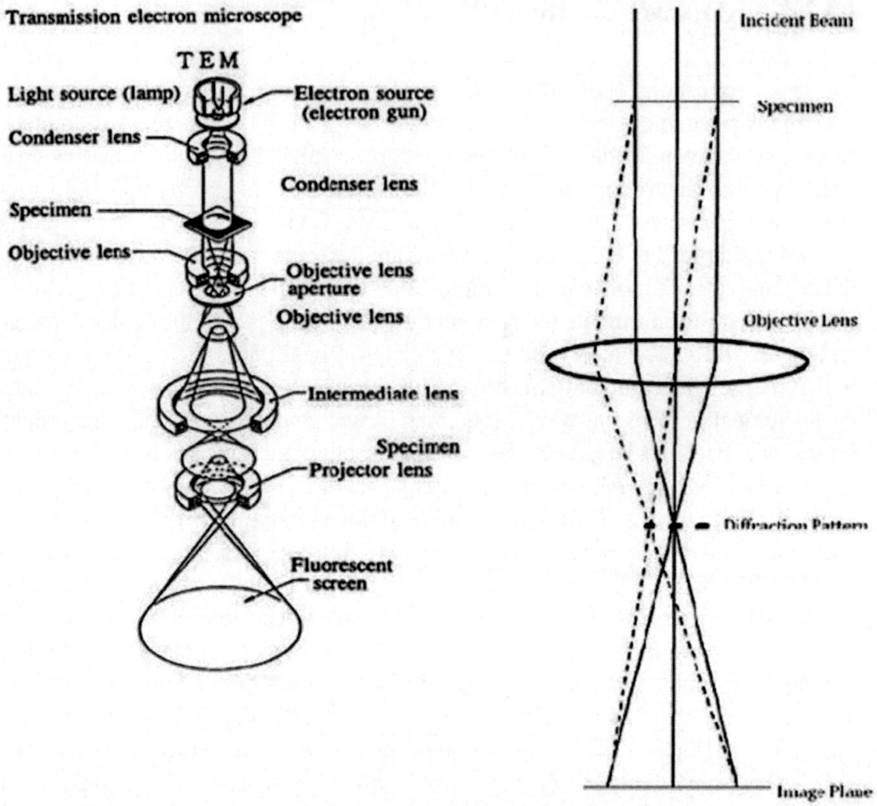

Fig. 5.4 General layout of a TEM describing the path of electron beam (left). A ray diagram for the diffraction mechanism in TEM (right).

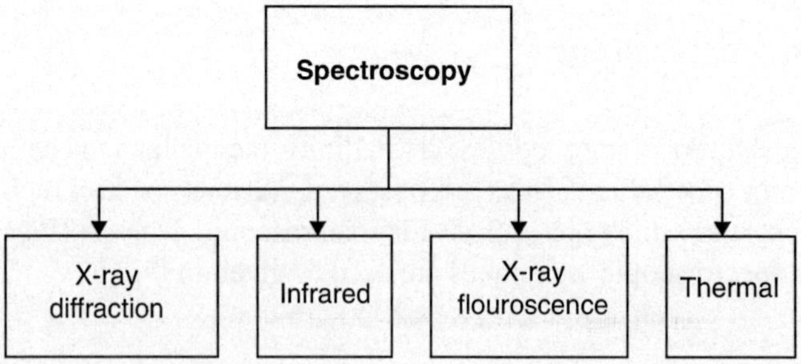

Fig. 5.5 Spectroscopic analytical methods.

X-rays in Analytical Methods

X-rays were discovered in 1895 by Wilhelm Conrad Röntgen, who received the first Nobel Prize in Physics in 1901 and are defined as having a wavelength of between

10-5 A to 100 A. In most applications x-rays are used between 0.1 and 25 A.

Uses of X-rays Include

Various analytical techniques have been developed utilising the properties of x-ray absorption, emission spectroscopy, x-ray diffractrometry etc (Table 5.1).

X-ray Analytical Methods

- **x-ray crystallography** in which the pattern produced by the diffraction of x-rays through the closely spaced lattice of atoms in a crystal is recorded and then analyzed to reveal the nature of that lattice.
- **x-ray astronomy** is an observational branch of astronomy, which deals with the study of x-ray emission from celestial objects.
- **x-ray microscopic analysis** uses electromagnetic radiation in the soft x-ray band to produce images of very small objects.
- **x-ray Fluorescenece** identifies the element from major to trace quantities in a sample.

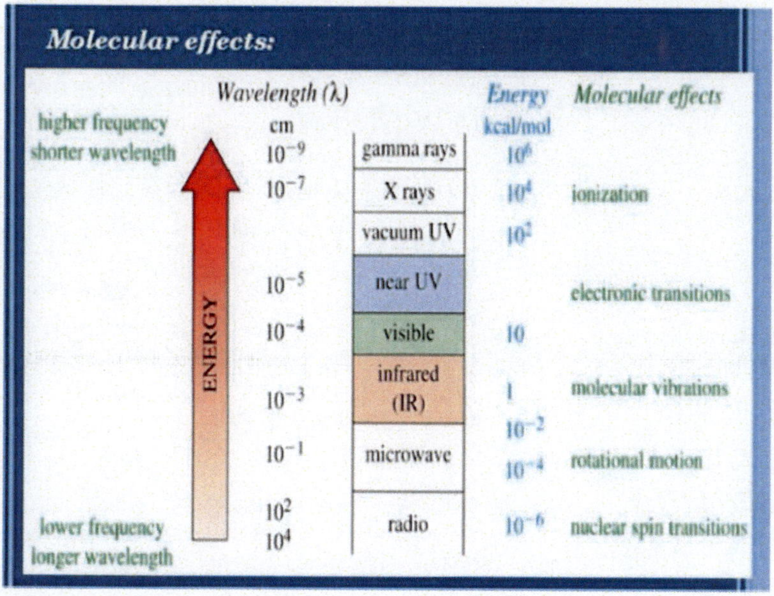

Fig. 5.6 Types of interactions with the variations of wavelengths.

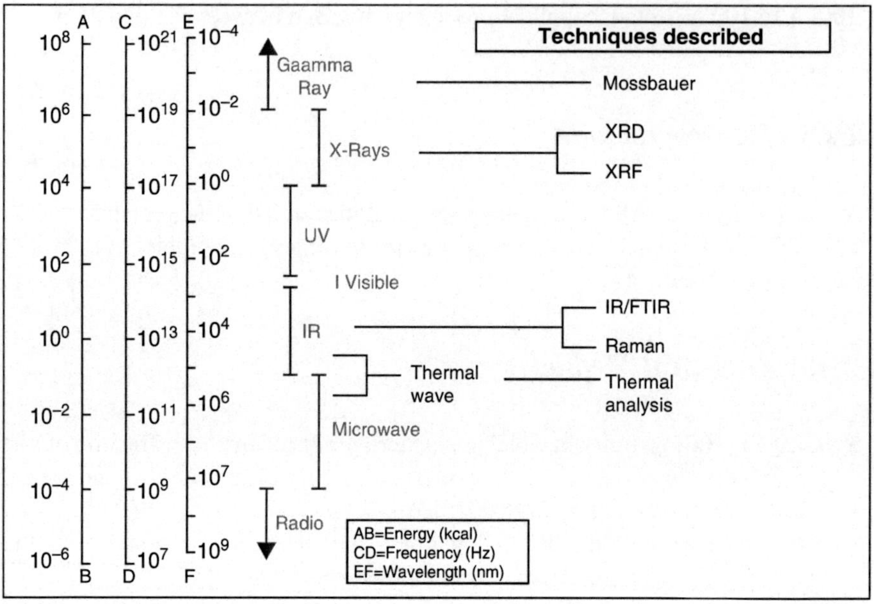

Fig. 5.7 Spectroscopic techniques used in different wavelengths.

Table 5.1 Different uses of x-rays and their respective photon energies

Use		**Accelerating potential**	**Target**	*Source type*	**Average photon energy**
x-ray crystallography		40 kV 60 kV	Copper molybdenum	Tube	8–17 keV
Diagnostic x-rays	Mammography	26–30 kV	Rhodium molybdenum	Tube	20 keV
	Dental	60 kV	Tungsten	Tube	30 keV
	General	50–140 kV	Tungsten	Tube	40 keV
	CT	80–140 kV	Tungsten	Tube	60 keV
Baggage screening	Carry-on/ Checked bags	80–160 kV	Tungsten	Tube	80 keV
	Container screening	450 kV–20 MV	Tungsten	Tube/ Linear accelerator	150 KeV–9 MeV
Structural analysis		150–450 kV	Tungsten	Tube	100 keV
x-ray therapy		10–25 MV	Tungsten/Hi ghZ material	Linear accelerator	3–10 MeV

Medical Uses

X-ray Transitions

Electron transitions to lower atomic levels in heavy atoms have quantum energies which place them in the x-ray region of the electromagnetic spectrum. The x-ray emissions associated with these transitions are called characteristic x-rays. Labels in Fig. 5.8 show the historical labelling of characteristic x-ray transitions. A typical spectrum of such transition is shown in Fig. 5.9 and for different metals, the frequency/spectrum variations is shown in Table 5.2.

Analyzing minerals is a very important task that involves several field of sciences because minerals possess unique features such as **electromagnetic spectra, and physico-chemical characteristics that distinguishes minerals from each other. The common methods used to analyze minerals are:**

- **XRD**
- **XRF**
- **Thermal Analysis**
- **Infrared Spectroscopy**
- **Electron Microscopy**

XRD

X-ray diffraction (XRD) is a nondestructive method to analyze crystalline phases of solid samples such as metals, ceramics, geological samples, electronic materials etc. and thereby finding out information on chemical composition. The sample is normally finely powdered before use.

Crystalline substances act as three-dimensional diffraction gratings for x-ray wavelengths similar to the spacing of planes in a crystal lattice. The typical interatomic spacing is ≈ 2–3 Å in crystals so the suitable radiation is x-rays. Hence, x-rays can be used for the study of crystal structures.

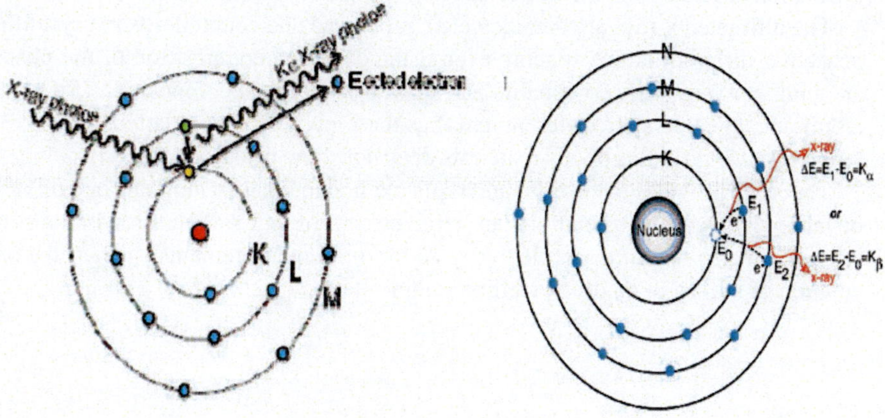

Fig. 5.8 The method of X-ray transition.

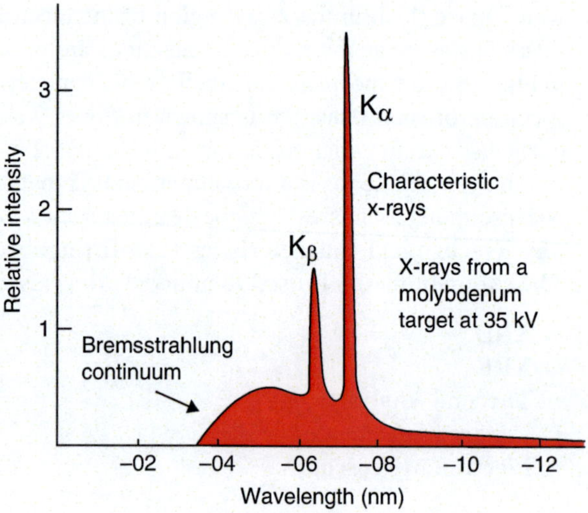

Fig. 5.9 K series X-ray pattern of an element.

Table 5.2 Values

Element	K_α *(weight average)*	$K_{\alpha 2}$ *(Strong)*	$K_{\alpha 1}$*(very strong)*	K_β *(weak)*
Cr	0.229100	0.229361	0.228970	0.208487
Fe	0.193736	0.193998	0.193604	0.175661
Co	0.179026	0.179285	0.178897	0.162079
Cu	0.154184	0.154439	0.154056	0.139222
Mo	0.071073	0.071359	0.070930	0.063229

x-rays are generated by a x-ray tube, filtered to produce monochromatic radiation, collimated (parallel rays of light) to concentrate, and directed towards the sample (Fig. 5.10). The interaction of the incident rays with the sample produces constructive interference and a diffracted ray when conditions satisfy Bragg's Law ($n\lambda = 2d \sin \theta$). that relates the wavelength of electromagnetic radiation (λ) to the diffraction angle (θ) and the lattice spacing (d) in a crystalline sample.

The diffracted x-rays are then detected, processed and counted. Every crystalline phase has different lattice spacing even if the chemical composition of the phases are similar, e.g. quartz, cristobalite and glass has chemical composition (SiO_2) but atomic arrangements are different and thus the x-ray diffraction pattern is also different (Fig. 5.11). Amorphous materials does not show peaks (glass).

The XRD provides a powerful and rapid method to identify minerals or a mixture of minerals. Data interpretations are straightforward and sample preparations are easy. It can also measure sample purity and textural measurements, such as the orientation of grains, in a polycrystalline sample.

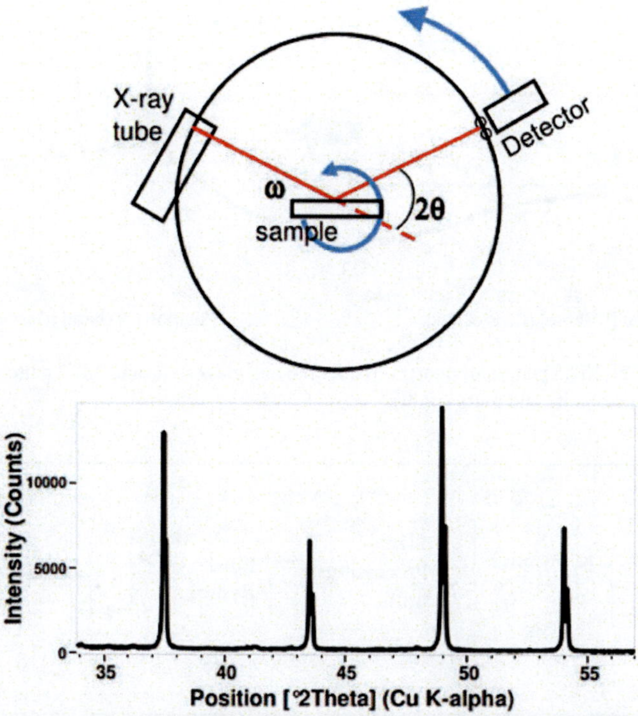

Fig. 5.10 Fundamental principle of XRD.

The X-ray Powder Diffraction Technique in Mineral Studies

In many cases, raw data needs processing (Fig. 5.12), i.e. offset and scaling, normalization, curve fitting, data smoothing, Fourier transform and a variety of other techniques.

Interpretation of data produces knowledge and is useful for making decisions. Decisions lead to actions that affect the process.

For Quality control the results may be used to alter production parameters, i.e. test of raw materials to choose suitable samples.

Charecteristics of information are: (i) reliability: the data is to be reproducible under the conditions used for the first experiment—a good quality data; (ii) data must be relevant for solving the problem under study and (iii) It must be responsive, i.e. should be available within time limit.

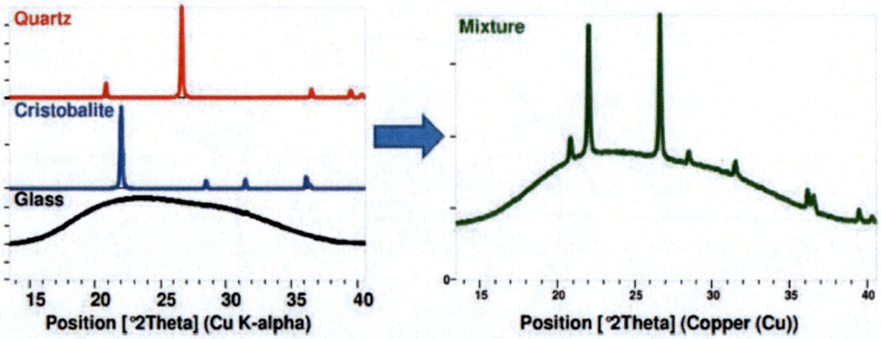

Fig. 5.11 The XRD pattern of quartz, cristobalite and glass (left) and XRD pattern of the mixture of quartz, cristobalite and glass (right).

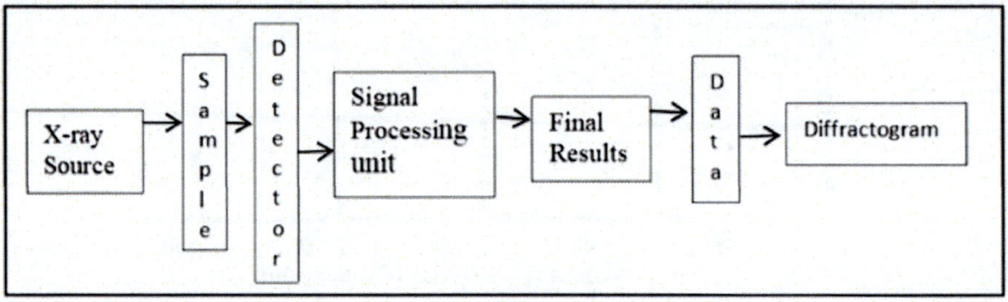

Fig. 5.12 Flow chart of x-ray diffraction data collection.

XRF

The x-ray fluorescence (XRF) is relatively non-destructive method for chemical and elemental analysis of rocks, minerals, sediments and fluids which is based on wavelength-dispersive spectroscopic principles that are similar to an electron microprobe.

When atoms of a material are exposed to high-energy, short wavelength radiation (e.g., x-rays), they become ionized. When the energy of the radiation is greater than the ionization energy, it is sufficient to dislodge a tightly-held inner electron. The atom becomes unstable and an outer electron replaces the missing inner electron. Due to this, energy is released because of the decrease in binding energy of the inner electron orbital compared with an outer one (electron energy in the inner shell is more negative). The emitted radiation is of lower energy than the primary incident x-rays and is termed fluorescent radiation. Every material emits radiation, which has energy characteristic of the atoms present (Fig. 5.13).

XRF is mainly used in chemical analysis of major and trace elements in rocks and sediments.

THERMAL ANALYSIS

Thermal analysis studies the properties of materials that change with temperature. Several methods commonly used are distinguished from one another by the property which is measured and is given in Table 5.3.

Thermal analysis is done by an instrument where the sample is placed inside a compartment that controls temperature and can measure properties of the sample. The data collection and processing are then done by the computer connected to it. Thermal analysis recognize, monitor and measures the changes due to thermal effects. When a material is heated its structural and chemical composition can undergo changes, e.g. fusion, melting, crystallization, oxidation, decomposition, transition, expansion and sintering. Thermal analysis techniques find broad application for the characterization of materials.

Application of Thermal Analysis

Following are the applications:

- DSC, TGA and DTA are often used for characterization and analysis of pharma materials.
- DSC is used for identification of various polymorphic transformation and structures.
- TGA is often used to measure residual solvents and moisture.
- Soil and clay analysis.
- Determination and compositional effects on glass transition.

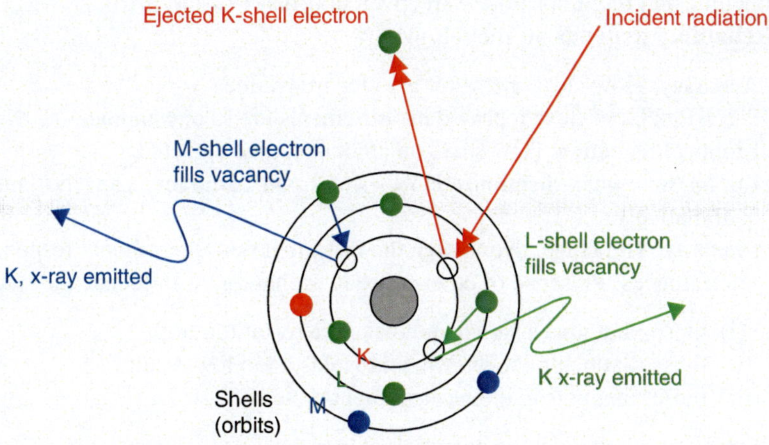

Fig. 5.13 Fundamental principle of x-ray fluorescence.

Table 5.3 Different methods of thermal analysis and theproperties of materials measured

Method	*Property measured*
Dielectric thermal analysis (DETA)	Pielectric permittivity and loss factor
Differential thermal analysis (DTA)	Temperature difference
Differential scanning calorimetry (DSC)	Heat flow changes
Dilatometry	Volume
Dynamic mechanical analysis (DMA)	Storage modulus (stiffness) and loss modulus (damping)
Evolved gas analysis (EGA)	Gases evolved during heating of a material, usually decomposition products
Laser flash analysis	Thermal diffusivity and thermal conductivity
Thermogravimetric analysis (TGA)	Mass change
Thermomechanical analysis (TMA)	Dimensional changes
Thermo-optical analysis (TOA)	Optical properties

IR-SPECTROMETRY

Energy in the form of infrared radiation when applied causes the vibration between the atoms of the molecules and when applied infrared frequency is equal to the natural frequency of vibration, absorption of IR radiation takes place and a peak is observed (Fig. 5.14).

Different functional groups absorb characteristic frequencies of IR radiation. Hence, gives the characteristic peak values (Table. 5.4). Therefore, IR spectrum of chemical substance is a fingerprint of a molecule for its identification.

General IR spectra of common clay minerals such as chlorite, illite, kaolinite, vermiculite and montmorillonite are given in (Figs 5.15 and 5.16).

Reliability depends on the following:

(a) Accuracy: How close the result is to the true value.
(b) Precision: How close repeated measurements are to one another.
(c) Limits of detection: How small an amount can be detected.
(d) Limits of quantification: How small an amount can be measured quantitatively.
(e) Linearity: The range, over which the instrument produces linear response.
(f) Interferences: Presence of other substances that might influence the test, e.g.:

    (i) those, that give a signal and make the result too high.
    (ii) those that tie up the analyte and yields a too low result.
    (iii) those, that poison the measurement system.

(g) Selectivity: how well the potential interferences are rejected.

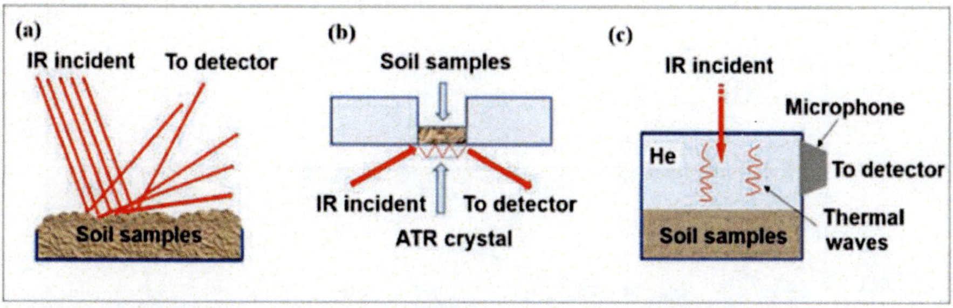

Fig. 5.14 Light path of: (a) diffuse reflectance spectroscopy (DRF), (b) attenuated total reflectance spectroscopy (ATR) and (c) photoacoustic spectroscopy (PAS).

Table 5.4 Properties of soil measured by different IR spectral region

Ir spectral region	*Spectral range*	*Soil properties and information*
Visible near infra red (VS-NIR)	400–1000 nm	Soil colour, iron content and composition, soil water, and organic matter
Acid exchange (cmol/kg) VS-NIR 400–2498 nm by Chang et al. (2001)		
Near infra red (NIR) or Short wave infra red (SWIR)	1000–2500 nm	Phyllosilicates, most sorosilicates, hydroxides, some sulphates, amphiboles, carbonates, soil water, and organic matter
Inorganic C (g/kg) NIR 1100–2498 nm by McCarty et al. (2002)		
Mid infrared (MIR) or thermal infrared (TIR)	2500–25000 nm	Quartz, feldspars, silicate minerals, mafic, clay, carbonate mineral group, and organic compounds
Al exchange (cmol/kg) MIR 2500–25000 nm by Janik et al. (1998) *Inorganic C (g/kg) MIR 2500–25000 nm by McCarty et al.* (2002)		

Standard material has a known value for some measurable parameter. Validation, calibration and quality assurance (QA) establishes the correctness of any laboratory result, which depends on standard material.

Calibration standard calibrates (or adjust) the quantification of analysis (or instrumental operation condition). It may use a zero concentration (or blank) to set the zero end of the scale and one more standard to establish a calibration range.

QA standard material is used for routine instrument checks and to validate that test method. Commercial standards are available with name, composition, catalog number, lot number, etc. In-house standard preparation can also be developed when needed.

CONCLUSION

Soil minerals study is very important in terms of investigating the major soil forming compounds. It also helps in finding out the fate of minor and trace elements, essential for soil-plant interaction. XRD technique is used to search out the phases for different types of samples whereas XRF is used for chemical and elemental

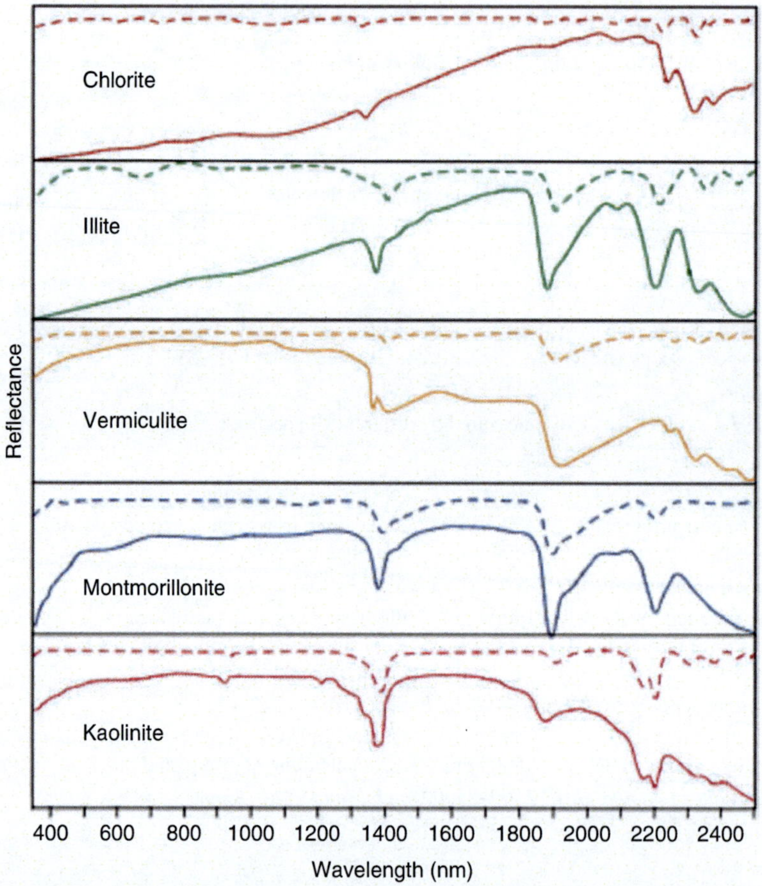

Fig. 5.15 Raw and continuum-removed spectra of five common clay minerals.

analysis of rocks and minerals. To study the property of materials with change in temperature, thermal analysis is used. Infrared spectroscopy can also measure different properties of soil (biological, physical and chemical). This is done by interaction of infrared radiation of different frequencies with matter.

THINK FOR A WHILE

- Why x-rays are used to study crystal lattices?
- How do we identify an amorphous material in a XRD?
- What kind of method is used to understand polymorphic structures?
- What is the difference between SEM and TEM?
- Why IR spectrum is called the fingerprint of a molecule?
- What are the factors that depend on standard material?

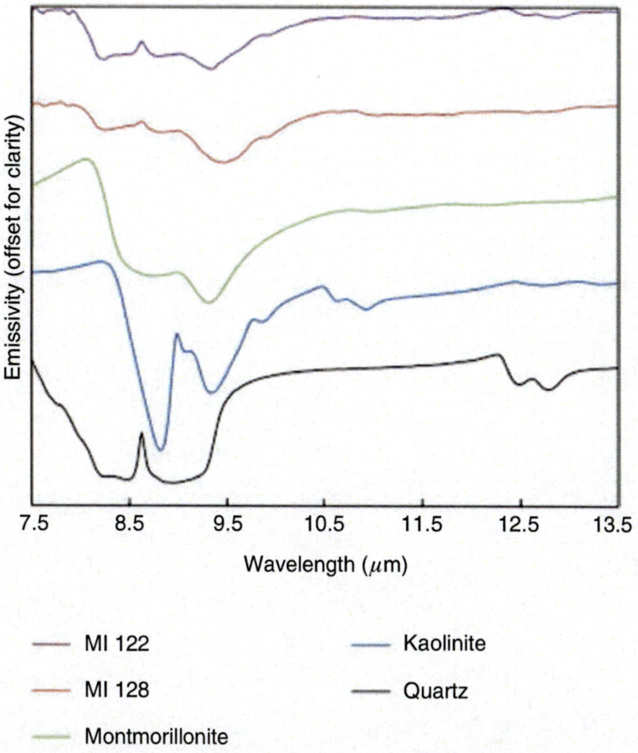

Fig. 5.16 Example MicroFTIR emissivity spectra of Tick Hill soil samples and ASL highlighting the presence of mixtures of clay and quartz minerals within the samples.

REFERENCES

Chang, Cheng-Wen, Laird, David, Mausbach, Maurice and Hurburgh, Charles (2001). Near-Infrared Reflectance Spectroscopy–Principal Components Regression Analyses of Soil Properties. *Soil Science Society of America Journal.* **65:** 480–490.

Ford, B.J. (2002). Using the Digital Microscope. Rothay House.

Janik, L., Skjemstad, J. and Merry, Richard (1998). Can infrared diffuse reflectance analysis replace soil extractions? Aust J Exp Agric. *Australian Journal of Experimental Agriculture – AUST J EXP AGR.* 38.

McCarty, G., Reeves, J., Reeves, V., Follett, Ronald and Kimble, J. (2002). Mid-Infrared and Near-Infrared Diffuse Reflectance Spectroscopy for Soil Carbon Measurement. *Soil Science Society of America Journal,* **66:** 640–646.

Part II
Soil Physics

Chapter 6
Physical Properties of Soils

Soil is made up of organic matter and mineral particles of different sizes. These are arranged in a matrix which results in pore spaces occupied by air and water. Different arrangements produce a three-phase system consisting of solids, liquids and gases. Soil consists of many physical properties such as size, shape, specific surface area, particle and bulk density, and porosity. These properties of soil greatly influence its uses in different domains.

SOIL COMPONENTS AND PHASES

Soil consists of four components and three phases. Soil components are inorganic matter, organic matter, water and air (Table 6.1). Inorganic matter consists of mineral components which are again divided into primary and secondary minerals whereas organic matter consists of components derived from plants and animals. The three phases of soil are solid, liquid and gas phase. Solid-phase of soil includes mineral and organic matter whereas the aqueous solution of same is its liquid phase. Gas phase consists of soil air which includes a mixture of gases such as O_2, N, CO_2, CH_4, etc. A typical soil consists of approximately 45% mineral matter, 5% organic matter, 20–30% water, and 20–30% air.

INORGANIC COMPONENTS

The inorganic component consists of 90–95% of the total solid fraction in soil. They are the result of weathering of parent materials. Inorganic components are again classified into primary and secondary minerals. Primary minerals are formed from igneous rocks through crystallisation whereas the secondary minerals are formed due to weathering of primary rocks and chemical alterations. Different properties of

S. Mukherjee, *Current Topics in Soil Science*,
https://doi.org/10.1007/978-3-030-92669-4_6

Table 6.1 Properties of different phases and components

Phases	*Solid*		*Liquid*	*Gas*
Components	*Inorganic*	*Organic*	*Soil solution*	*Soil air*
Composition	Primary minerals from weathering such as quartz, feldspar, garnet, hornblende, silicates, secondary minerals.	Plant and animal remnants; living organisms, usually <5%	Aqueous solution of ions (e.g., Na, K, Ca, Mg, Cl, NO_3, PO_4, SO_4)	N_2, O_2, CO_2, CH_4, C_2H_6, H_2S, N_2O, NO
Properties	Skeleton matrix ρ_s = 2.0–2.8 Mg/m^3	Large surface area, very active, affects CO_2 in the atmosphere ρ_s = 1.2–1.5 Mg/m^3	Heterogeneous, dynamic, discontinuous ρ_w = 1.0 Mg/m^3 w	ρ_a = 1–1.5 kg/m^3 variable, dynamic

Note: ρ_s = particle density, ρ_w = density of H_2O, ρ_a = density of air

inorganic components contribute to different types of soil. These properties are particle size, shape, texture, colour, surface area, bulk density, packing arrangements, the composition of clay minerals, etc.

PARTICLE SIZE

In inorganic components, two types of mineral particles occur – primary and secondary particles. Primary particles are individual discreet units whereas secondary particles are the aggregate of primary particles i.e. two or more primary particles consolidate to form secondary particles. Particle size is measured by its effective diameter. It is defined as the diameter of a sphere that has a velocity of fall in a liquid medium equal to the particle in question. It is a very important property of soil as its affects other properties like texture, porosity, surface area, etc.

SIZE FRACTIONATION

Primary particles are mainly divided into sand, silt and clay depending on their size. There are numerous systems of classifying separates into different size classes. The most common system used is the U.S. Department of Agriculture (USDA) and the International Society of Soil Science (ISSS) (Table 6.2).

Particles >2 mm are considered the non-soil fraction as they do not affect the fertility and productivity of the soil. However, tilling practices may get affected in soils with greater non-soil fractions. The size of these particles are: pebbles (2 mm–7.5 cm), cobbles (7.5–25 cm), stones (25–60 cm), and boulders (>60 cm). The general physical properties of sand, silt and clay are listed in Table 6.3.

Table 6.2 Different particle sizes in different systems of classification

Particle	*USDA (mm)*	*ISSS (mm)*
Gravel	>2	>2
Very coarse sand	1–2	
Coarse sand	0.5–1.0	0.2–2.0
Medium sand	0.25–0.50	
Fine sand	0.10–0.25	0.02–0.20
Very fine sand	0.05–0.10	
Silt	0.002–0.05	0.002–0.020
Clay	<0.002	<0.002

Table 6.3 General physical properties of sand, silt and clay

Properties		*Soil separates*	
	Sand	Silt	Clay
Size (mm)	2–0.02	0.02–0.002	<0.002
Shape	Jagged	Slightly irregular	Platy/tube-like
Feel	Gritty	Smooth, floury	Sticky
Plasticity	Non plastic	Slightly plastic	Plastic
Cohesion	Non cohesive	Slightly cohesive	Cohesive, gelatinous
Surface area	Very low	Moderate	Very high
Mineralogy	Primary	Primary minerals	Secondary clay minerals
Heat of wetting	None	Minimal	High
Secondary Particles	None	Few	Forms aggregate
Water holding capacity	None/slight	Moderate	High, hygroscopic
Hardness (Mohs scale)	5.5–7	5.5–7.0	–
Ion exchange capacity	None	Very low	High to very high

ASSESSMENT OF PARTICLE SIZE FRACTION

To find out the size of soil particles, two methods are used, namely *dispersion* and *fractionation*. Dispersion is the removal of cementing materials from the aggregates to break them into primary particles. For this, several cementing agents are used which depends on the type of cementing materials. For example, hydrogen peroxide (H_2O_2) is used to remove organic material, dilute acid to remove carbonates/electrolytes, and sodium dithionite to remove sesquioxides (oxide containing three atoms of oxygen with two atoms of another element). Fractionation is the process by which soil particles are physically separated into different size ranges with the help of sieving or sedimentation. Sieving is the most common process of fractionation in which the wire mesh used are of different mesh sizes to separate the particles into particular sizes (Table 6.4).

Table 6.4 Different types of fractionation methods used for different particle sizes

Methods of fractionation	*Approximate size range (mm)*
Sieving	100–0.5
Sedimentation	2 < 0.002
Gravity sedimentation	0.1–0.0005
Centrifugal sedimentation	0.1–0.00005
Permeability	0.1–0.0001
Optical microscope	1.0–0.001
Electron microscope	0.005–0.00001
Elutriation	0.05–0.005
Gas absorption	0.1–0.0001
Turbidimetry	0.005–0.00005

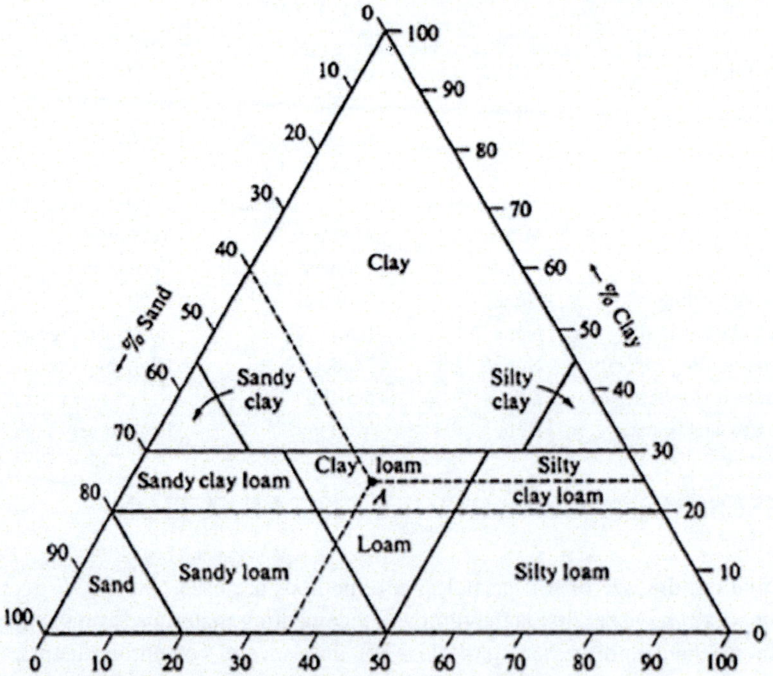

Fig. 6.1 Textural triangle.

INTERPRETATION OF RESULTS

Numerous methods are used to interpret the results of particle size analyses. The data are commonly expressed as one of the procedures like textural classes, summation curve, uniformity coefficient, etc. but the most common and simplest procedure used is the textural triangle (Fig. 6.1).

The textural triangle refers to the relative proportion of sand, silt and clay in soil. Each class in the soil is based on these relative proportions. For example, soil with a greater proportion of sand is called sandy soil whereas soil with an equal proportion of sand, silt and clay is called loam. Loam is then divided into silty loam, sandy loam and clay loam depending on the particle size percentage.

PARTICLE SHAPE

Like particle size, the shape is also an important property of soil texture. It mainly depends on the nature of the parent material and its degree of weathering. It influences specific surface area and particle size analysis as well as the packing of particles and soil strength. Particle shape is defined by two aspects i.e. *roundness* and *sphericity* (Fig. 6.2). Roundness refers to the degree of sharpness of the edges of a grain. The smoother the edges, the more is its roundness. It is computed as follows:

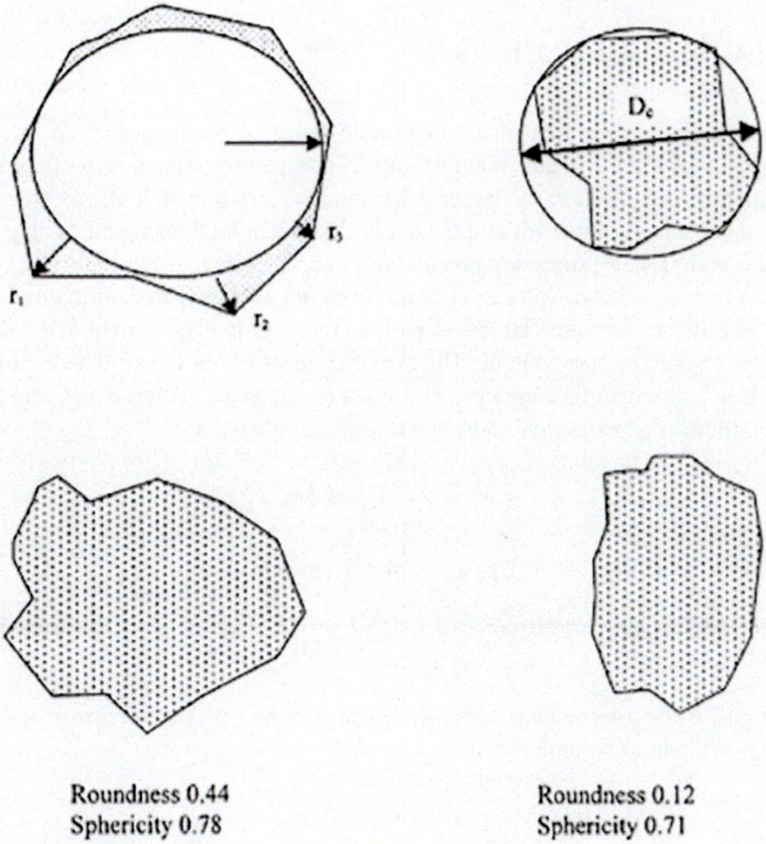

Fig. 6.2 Roundness and sphericity as expressions of particle shapes.

Roundness =

where r_i is the radius of the curvature of the edges of a grain, R is the radius of the largest circle inscribed within the particle, and n is the number of corners in a particle.

Sphericity or form refers to the overall configuration of a particle and it depends on the relative lengths of the three major axes (short, intermediate and long). If all the axes are of the same length, then it is a perfect sphere. It is computed as:

$$\text{Sphericity} = D_d / D_c$$

where D_d is the diameter of a circle with an area equal to that of the particle projection as it rests on its flat side, and D_c is the diameter of the smallest circumscribing circle.

Both sphericity and roundness have values between 0 and 1. A grain with perfect roundness and sphericity has values close to 1.

SPECIFIC SURFACE AREA

The specific surface area of a soil sample refers to the total surface area of the sample contained in a unit mass of soil. It is inversely related to particle size i.e. smaller the particle size, the larger is its specific surface area. It affects soil's physical and chemical properties and is largely dependent on the amount of clay present in the soil. These properties are texture, cation exchange capacity (CEC), fluid adsorption, plasticity, water retention, cohesion, saturated hydraulic conductivity, etc. Specific surface area increases with an increase in clay content and a decrease in soil organic matter content. The specific surface area is expressed using three separate indices: surface area per unit mass (a_m), per unit volume (a_v), and per unit bulk volume (a_b) as expressed by the following equations:

$$a_m = A_s / M_s \left(\text{m}^2/\text{g}\right)$$

$$a_v = A_{ss} / V\left(\text{m}^2/\text{m}^3\right)$$

$$a_b = A_{st} / V\left(\text{m}^2/\text{m}^3\right)$$

where A_s is the total surface area, M_s the mass of soil, V_s the volume of soil solids, and V_t is the total volume.

Table 6.5 Particle density of some common minerals

Mineral	*Particle density (Mg/m^3)*	*Mineral*	*Particle density (Mg/m^3)*
Biotite	2.7–3.3	Kaolinite	2.61–2.68
Brucite	2.38–3.40	Magnetite	5.175
Calcite	2.72–2.94	Montmorillonite	2.0–3.0
Chlorite	2.60–3.3	Muscovite	2.77–2.88
Diamond	3.50–3.53	Orthoclase	2.55–2.63
Dolomite	2.86	Pyrite	5.08
Gibbsite	2.38–2.42	Quartz	2.65
Goethite	3.3–4.3	Serpentine	2.55
Gypsum	2.3–2.47	Talc	2.58–2.83
Haematite	5.26	Tourmaline	3.03–3.25
Hornblende	3.02–3.45	Vermaculite	2.3
Illite	2.60–2.90		
Other constituents		*Particle density (Mg/m^3)*	
Soil organic matter		1.0–1.4	
Water		1.0	
Air		10×10^{-3}	

Source: Adapted from Handbook of Chemistry and Physics (1988)

PARTICLE DENSITY

Particle density is defined as the weight of a particle per unit volume. It is expressed in grams per cubic centimeter (g/cm^3). The average particle density for mineral soils is usually given as 2.65 g/cm^3 because of quartz which is a predominant soil mineral (Table 6.5). Particle density depends on the relative proportion of minerals and organic matter in the soil. It determines other physical properties of soil such as bulk density and porosity.

BULK DENSITY AND POROSITY

In order to quantify the state of compaction and the amount of pore space in soil, the volume and mass of solids are related to bulk density. It is calculated as the mass of soil (M_s) divided by its volume (V_t). The bulk density of soil is mainly calculated in the dry state (ρ_s) i.e. soil without water but the calculation of wet density (ρ'_s) is needed in some experiments where the mass of water (M_w) is included. Wet bulk density changes with time due to evaporation thus not preferred. It is expressed as g/cm^3 or Mg/m^3.

$$s =$$

$$Vt =$$

Bulk density depends on several factors such as soil texture, compaction, porosity, organic matter content, etc. Soil with greater compaction contains less porous space thus bulk density increases. Soils rich in organic matter have lower bulk density. Most of the soils have bulk densities between 1.0 and 2.0 Mg m^{-3}.

POROSITY (*F*)

Porosity is defined as the part of the bulk volume of soil that does not contain any solid material. The void spaces are occupied by air or water or both. Depending on the structure and texture of the soil, pore spaces can be large (macropores) or small (micropores). The function of macropores is to facilitate exchanges of fluids whereas micropores hold water through surface tension (Fig. 6.3). Soil porosity can be expressed in four ways:

1. **Total porosity** (f_t): It is defined as the ratio of the volume of fluids (water and/or air) (V_f) to total volume(V_t) of soil.

$$f_t = \quad V_f / V_t = V_g + V_l / \left(V_s + V_l + V_g\right)$$

2. **Air-filled porosity** (f_a): It is defined as the ratio of the volume of air-filled pores to the total volume. Air porosity is also equal to the difference between total porosity and the volumetric moisture content.

$$f_a = \quad V_g / V_t$$

3. **Void ratio** (*e*): It is defined as the ratio of the volume of voids to the volume of solids.

$$e = \quad V_f / V_s = \left(V_g + V_l\right) / \left(V_{in} + V_o\right)$$

4. **Air ratio** (α): It is defined as the ratio of the volume of air to that of the solids

$$\alpha = \quad V_g / V_s = V_{in} / \left(V_o + V\right)$$

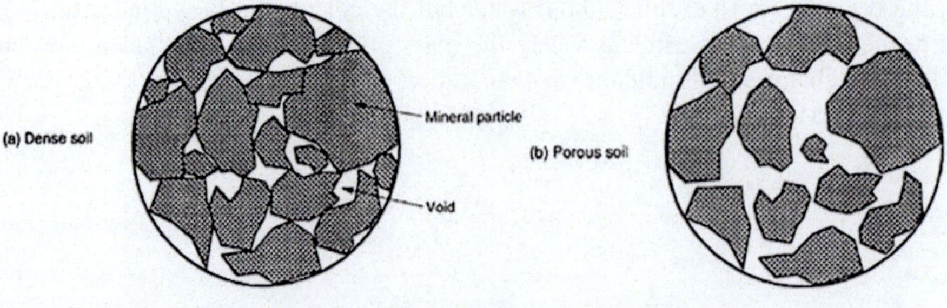

Fig. 6.3 Porosity in different types of soil: (a) dense soil and (b) porous soil.

BASIC MECHANICS OF SOIL

Atterberg Limits

Soil can be remoulded without crumbling in presence of clay minerals and moisture. This is due to the cohesive nature of clay minerals by which it adsorbs water. In 1911, a Swedish scientist named Albert Atterberg developed a method to describe the consistency of fine-grained soil with varying moisture content. With varying moisture content, the consistency of soil is divided into four states, viz. *solid, semi-solid, plastic* and *liquid* (Fig. 6.4).

Shrinkage Limit The moisture content in percent, at which transition from solid to semi-solid takes place. It is the maximum limit of water content where the further loss in moisture will not reduce the soil volume. It is less frequently used than plastic and liquid limit.

Plastic Limit The soil moisture content at which any increase in the moisture content will cause a semi-solid soil to become plastic.

Plasticity Index is calculated as the difference between the liquid limit and the plastic limit. The larger the value of the plasticity index, the greater is the plasticity of the soil. High plasticity index indicates a high amount of clay and the soil is highly compressible.

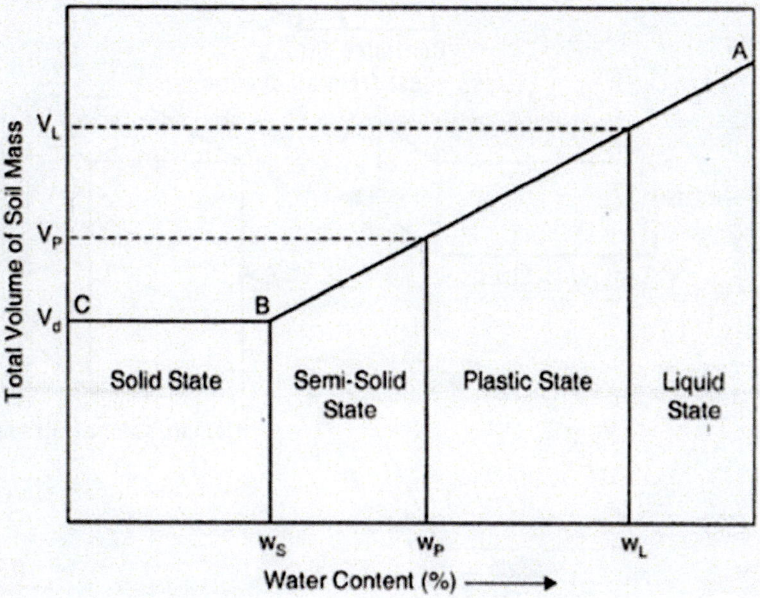

Fig. 6.4 Different Atterberg limits with the change of moisture content. w_S, w_P, w_L are the shrinkage, plastic and liquid limits, respectively.

Liquid Limit The water content in the soil that causes soil to change from liquid state to plastic state. It is also called the upper plastic limit.

Liquidity Index is used to indicate the consistency of undisturbed soils. It is determined by subtracting the Plastic limit from the natural water content of the sample and then dividing it by the plasticity index.

The three phases of soil (solid, liquid and gas) interacts with each other to form complex states of stress and strain. These stresses and strains are studied carefully and quantified to be used in the construction and design of bridges, tunnels, etc.

Stress It is defined as the force acting on a body per unit area. This force causes a body to deform. Its dimension is $ML^{-1} T^{-2}$.

Types of Stress

There are three types of stress (Fig. 6.5). They are:

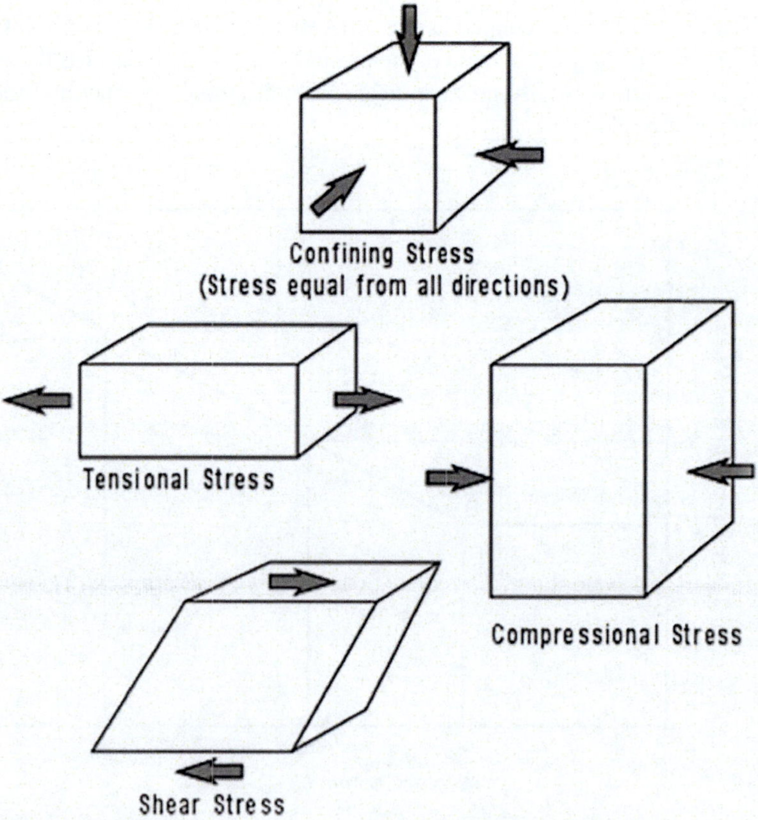

Fig. 6.5 Types of stress.

Normal stress: It is the stress that is applied normally or perpendicular to the cross-sectional area of the body. It is of two types: tensional and compressional stress.

Shear stress: It is the restoring force or deforming force that acts parallel to the surface area.

Volumetric stress: When stress is applied in all three directions changing the volume of the body, it is called volumetric stress or bulk stress.

Strain: It is the ratio of change in size or shape to the original size or shape and is dimensionless.

Types of Strain

There are three types of strain (Fig. 6.6). They are:

Longitudinal strain: Longitudinal strain is the ratio between the change in length to its original length. It is divided into tensile and compressive strain.

Volumetric strain: It is the ratio of change in volume to its original volume under normal loading conditions.

Shear strain: Change in the shape of a body under shear stress is called the shear strain. It is the change in tangential angle in the direction of stress.

Stress Strain Curve

The stress strain curve describes the relationship between stress and strain. As the stress increases in a material, it passes through few stages of strain (Fig. 6.7):

Elastic deformation—The deformation in which the strain is reversible when stress is removed.

Ductile deformation—The deformation in which the strain is irreversible when stress is removed.

Fracture—The strain is irreversible until the material breaks.

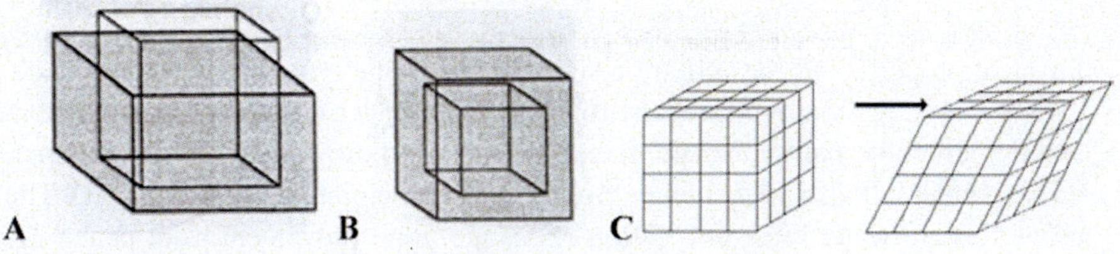

Fig. 6.6 Types of strain: A. Longitudinal, B. Volumetric and C. Shear.

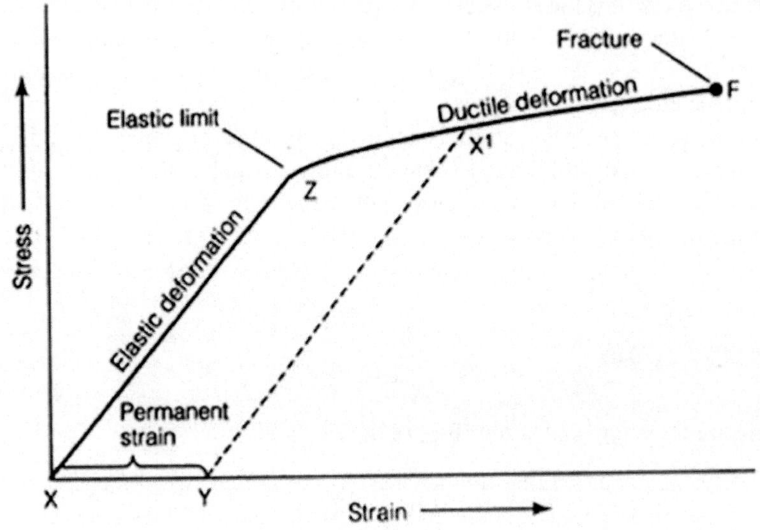

Fig. 6.7 Stress strain curve describing different types of deformation in rock.

FACTORS CONTROLLING STRESS STRAIN BEHAVIOUR OF SOILS

- Soil composition: Includes mineralogy, grain size and its distribution, the shape of particles, pore fluidity and its content and ions in pore fluid and grains.
- Soil structure: Arrangement of particles in the soil.
- Degree of saturation: Includes pore water pressure, permeability, total and effective stress, drained or undrained conditions.
- Loading conditions.
- Initial state: Includes normal and overconsolidated soil behaviour.
- Density/Confinement.

CONCLUSION

Physical properties of the soil are a significant parameter to determine its usage in diverse fields, be it agriculture, environment or engineering. The properties such as texture, structure, particle shape, size, bulk density, porosity, etc. not only depends on each other but also affects other biological and chemical properties of soil. These properties are water retention, cation exchange, fluid movement, root penetration, nutrient exchange, etc. Thus these properties should be thoroughly studied and measured for their use in agriculture and management.

THINK FOR AWHILE

- A soil body has a dimension of 10 cm × 10 cm × 10 cm. Its dry weight is 1200 g having a water content of 260 g. Find the bulk density, total porosity, air-filled porosity and specific surface area of the soil assuming particle density to be 2.65 mg m–3.
- The bulk density and particle density of soil are 1.5 and 2.5 g cm–3, respectively. If the moisture in the soil is 15%, then find the porosity, aeration porosity, degree of saturation and void ratio.
- What are different methods for the expression of results of particle size analyses?
- What is the difference between roundness and sphericity?
- What is the proportion of soil separates in sandy clay loamy soil?

Chapter 7
Soil Structure

The arrangement or organization of the particles in the soil (i.e., the internal configuration of the soil matrix) is called soil structure. This concept, however, requires a clear understanding of the word "particle". As far as structure is concerned, soil particles refer not only to the individual mechanical elements, such as sand, silt, and clay, but also to the aggregates or structural elements, which have been formed by the aggregation of smaller mechanical fractions. The word "particle", therefore, refers to any unit that is a part of the makeup of the soil, irrespective of its being a primary (sand, silt, or clay fraction) or a secondary (aggregate) particle. Consequently, the structure of a soil implies an arrangement of these primary and secondary particles into a certain structural pattern (Baver, 1956).

Soil texture refers to relative proportion of sand, silt, clay whereas soil structure refers to its arrangement into a definite pattern. Texture is a basic property of soil and cannot be changed easily but soil structure can be changed easily by many management practices like ploughing, liming, manuring etc.

CLASSIFICATION OF SOIL STRUCTURE

Complete classification of soil structure based on Nikiforoff (1941) include the: (1) type, which notes the shape and arrangement of particles (Table 7.1); (2) class, which indicates particle size (Table 7.2); and (3) grade, which indicates the durability or stability of the particles and their ease of separation (Table 7.3).

Basic soil types are discussed in Fig. 7.1 in which the following tables are described.

S. Mukherjee, *Current Topics in Soil Science*,
https://doi.org/10.1007/978-3-030-92669-4_7

Table 7.1 Types of soil structure

Type: Shape and Arrangement of particles	
Structure	
Platy	Aggregates with horizontal axes longer than the vertical
Prismatic	Aggregates with horizontal axes shorter than the vertical
Blocky	Aggregates with horizontal and vertical axes more or less equal
Granular	Aggregates that are more or less rounded

Table 7.2 Classes of soil structure

Class: Size of soil particles	
Very coarse	> 10 mm, except blocky type is > 5 cm.
Coarse	5–10 mm, except blocky type is 2–5 cm.
Medium	2–5 mm, except blocky type is 1–2 cm.
Fine	1–2 mm, except blocky type is 5–10 mm.
Very fine	< 1 mm, except blocky type is 1–5 mm.

Table 7.3 Grades of soil structure

Grade: Stability of particles
Poorly developed
Weakly developed
Moderately developed
Well developed
Strongly developed

GENESIS OF SOIL STRUCTURE

The genesis of soil structure refers to the causes and methods of formation of the structural units or aggregates. Soil structure is a result of the processes of *flocculation* and *granulation*. Flocculation is the aggregation of individual particles into small clumps or clusters. Granulation is the cementation of aggregates into masses (Fig. 7.2).

Flocculation and stable aggregate formation are not synonymous. The former is primarily electrokinetic in nature. Primary particles with a high electrokinetic (zeta) potential repel each other when they collide in a suspension. When the potential is lowered sufficiently, a collision between particles results in a mutual attraction and the formation of a floccule which is stable as long as the flocculating aggregate is present.

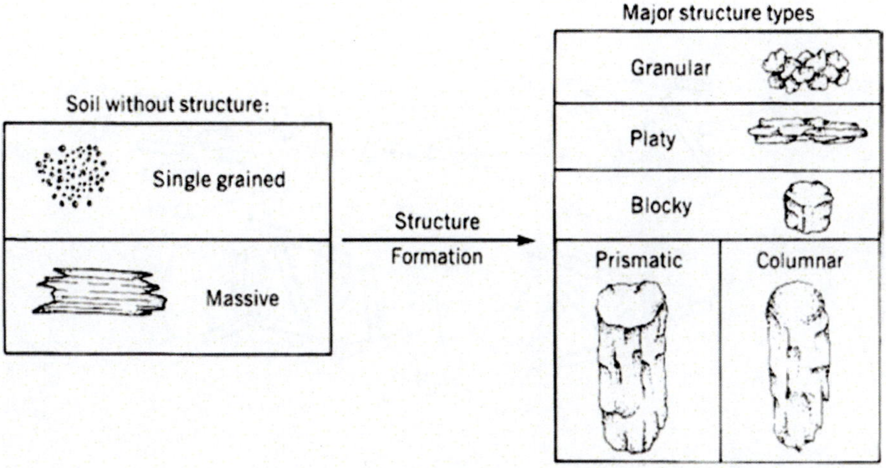

Fig. 7.1 Formation of various types of soil structure from single grained and massive soil conditions.

Stable aggregate formation requires the primary particles be firmly held together so that they do not disperse in water. In other words, from the point of view of soil structure, aggregate formation requires a cementation or binding together of flocculated particles. Thus, flocculation may aid in the aggregation process but is not aggregation in itself (Baver, 1956). As Bradfield (1936) so aptly states, "Granulation is flocculation plus!"

According to Kohnknke (1968) the factors responsible for aggregate formation are the: (a) amount and nature of clay minerals, (b) extent of flocculation, (c) organic-inorganic linkages, (d) microbial glue and (e) mineral cementing agents like iron and aluminium oxides. Combined effects of more than one factor have, however, been widely recognized.

Cation Effect on Soil Aggregation

It has been generally concluded that certain cation effect on soil aggregation helps in the improvement of soil structure. The influence of wide variety of exchangeable cation ratios (such as Ca, Mg, Fa, K and H) with and without addition of organic matter was determined by Martin and Aldrich (1955).

(A) The marked effect of calcium on the thixotropy of clay suspensions and on the water stability of clay films, particularly where pectin was present, was an evidence of the validity of idea of calcium linkage in the formation of water-stable soil aggregates (Peterson, 1947). Martin et al. (1966) observed that calcium saturation of the soil colloids contributes to the formation of stable aggregate and further noticed that poor structure of alkali soils could be modified into a

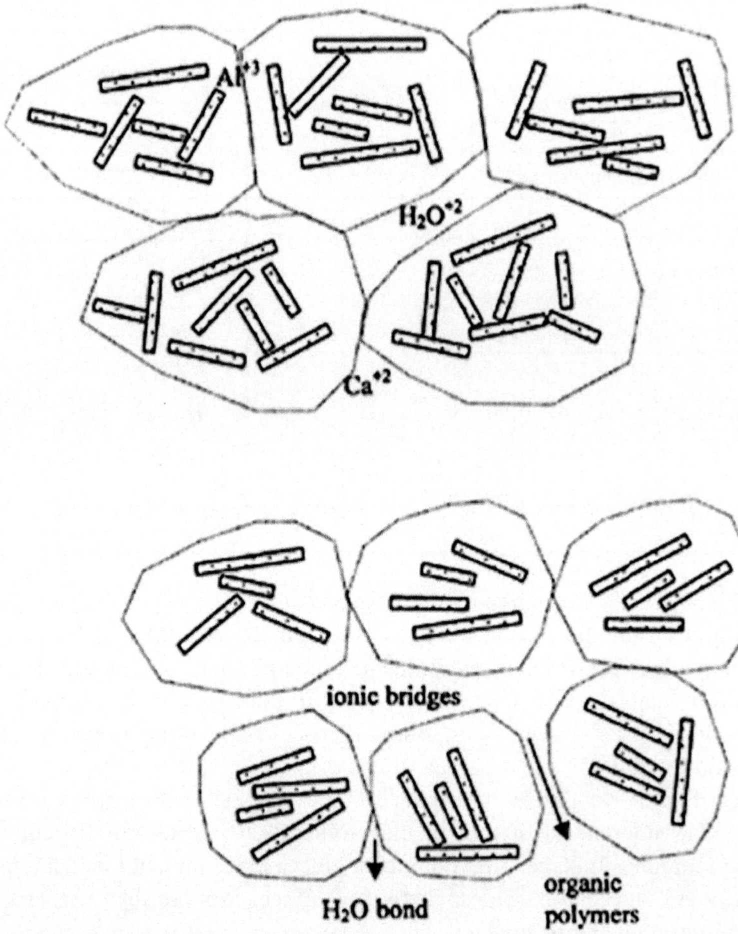

Fig. 7.2 Floccules of clay particles cemented together lead to granulation. Aggregation is floccul ation+cementation.

favourable physical condition, if the exchangeable sodium could be replaced by calcium by using gypsum and organic matter.

(B) High percentage of Mg^{2+} had no effect on aggregation.

(C) Depending on the soil, a high percentage of exchangeable K^+ may or may not reduce aggregation.

(D) Increase in Na^+ always greatly reduced aggregation.

(E) H^+ was found to have no influence whatsoever.

(F) An excess of free lime reduced aggregation.

(G) Organic matter additions increased aggregation, regardless of the exchangeable cation status of the soil though the effect was least in presence of high sodium salt.

Organic Matter and Aggregation

Organic matter plays tremendous influence on the formation of water-stable aggregates (Kononova, 1966; Martin, 1971). It is a well-observed fact that organic matter must be made soluble or oxidized to ensure complete dispersion of soils in mechanical analyses.

Baver (1956) observed that if various soils are grouped according to their clay contents, it is seen that the effect of organic matter is more pronounced in those soils containing the smaller amounts of clay (Table 7.4). Thus, clay and organic colloids seem to be responsible for the major portion of the soil aggregates. Organic matter, however, appears to be more effective in causing the finer fractions to aggregate, since the extent to which the silt and clay separates were aggregated showed a correlation higher with organic matter than clay. These results suggest that organic colloids not only cause a high degree of aggregation of the clay particles but also produce large aggregates (Fig. 7.3).

Inorganic Matter and Aggregation

1. **Clay Minerals**

The content and mineralogical composition of clay are two important factors that influence soil structure. To a certain degree, clay mineralogy influences such properties as soil stability, state of soil aggregation and permeability that controls directly or indirectly soil erosion (Biswas and Karale, 1974). Amount, quality and nature of clay minerals influences the total quantity of aggregates of a soil.

Peterson (1944) concluded that montmorillonite produced aggregates in which the horizontal and vertical axes are approximately equal whereas kaolinite forms a platy structure. H-saturated kaolinite produced more definite platiness than Ca-saturated kaolinite.

Table 7.4 Correlation of aggregation with clay and organic matter

Number of samples	*Carbon (%)*	*Correlation between aggregates larger than 0.05 mm and 5-μ clay*
18	<1	0.947(0.456 significant)
33	1–2	0.506(0.347 significant)
35	>2	0.376(0.396 significant)
Number of samples	*5-μ clay (%)*	*Correlation between aggregates larger than 0.05mm and organic matter (% carbon)*
15	>35	0.663(0.514 significant)
29	25–35	0.650(0.370 significant)
33	<25	0.758(0.30 significant)

Fig. 7.3 Schematic of the relationship between aggregate stability (mean weight diameter) and soil organic matter content for a group of soils (A, B, C) as already discussed above.

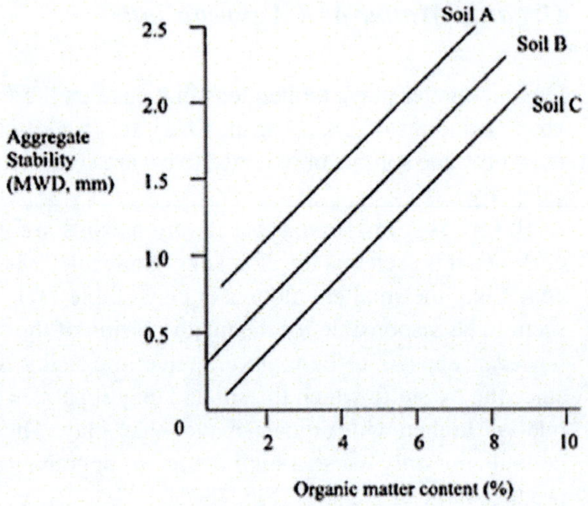

Azuma et al. (1968) reported that aggregation, particularly the formation of small aggregates, increased with increasing clay content. Clay particles themselves, by virtue of their properties of adhesion and cohesion, stick together to form aggregates (Daji, 1970). Sand and silt particles cannot form aggregates as they do not possess the power of adhesion and cohesion. As these particles carry a coating of colloidal clay, they are enmeshed in the aggregates formed by adhering clay particles.

Iron and Aluminium Oxides

There is a good relationship between aggregate formation and sesquioxides which form irreversible and slowly reversible colloids and thus help to form water stable aggregates.

From electron microscopic studies, Kroth and Page (1947) concluded that iron and aluminium oxides provided a continuous matrix which bound soil particles into secondary units by physical forces alone.

Filippovich (1956) found that appreciable improvement of structure would not be accomplished unless humus is accompanied by colloidal hydrated iron oxides.

Biswas et al. (1961) found a positive correlation between aggregation and free iron oxides in almost all soils of India except in saline soils of Delhi and Hyderabad. On the other hand, Deshpande et al. (1964) found that removal of iron oxides by selective chemical treatment did not reduce stability of aggregates from some soils high in iron oxides. They concluded that iron oxide was present as discrete crystals and did not cement particles.

Free iron oxide had negligible role on soil aggregation in lateritic and saline soil but a significant positive correlation was found between free iron oxide and aggregation in alluvial soil.

Climate and Aggregation

It is important to know not only the total percentage of aggregates in the soil but also the amount of silt and clay (in percentage) present in the form of aggregates. That is why course-textured soils do not have so much silt and clay to aggregate as the finer-textured ones.

Baver (1956) analyzed number of aggregates of different soils that has yielded significant datas for a correlation between climate and aggregation (Fig. 7.4). Several striking relationships exist between aggregation and rainfall, when temperature is kept constant (Fig. 7.4a). The percentage of aggregates in a given weight of soil is maximum in the semi-arid and semi-humid regions. This is where the chernozem-like and humid-prairie soils are located.

The percentage of aggregates is low in desert soils because of small clay content. This is due to low amount of chemical weathering. Low chemical weathering causes low amount of clay content which results in small number of aggregates to be present. As the rainfall becomes greater, chemical weathering is intensified, and clay formation also increases.

Not only is the clay content highest in the chernozem and dark humid-prairic regions but the percentage of organic matter in these soils is also largest. As a result of these two factors, the percentage of aggregates in the soil is greatest in the chernozem and dark humid-prairie soils.

It is seen that podsol soils have a relatively low perccntage of the silt and clay aggregated. The low extent of aggregation in the podsol type of soils is probably due to the low organic-matter content and the removal of alumina and iron from the A-horizons.

The larger percentage aggregation of the silt and clay in the regions of decreasing rainfall is due to organic matter and divalent bases. The presence of sodium in truly desert soils, together with low amounts of organic matter, causes only a slight aggregation of the small quantities of silt and clay in the soils.

When rainfall is kept constant and temperature is increased (Fig. 7.4b), the percentage of aggregation of the silt and clay is affected differently in humid and semi-arid regions. The chernozem-like soils of the semi-arid regions, the percentage aggregation of the silt and clay decreases. This is due to the decrease in aggregation due to a lowering of the organic-matter content with the increasing temperature.

These tendencies indicate that the soil-forming climatic factors of rainfall and temperature are expressed in the structure of the major soil groups. In addition to these broader effects of climate, there are also the influences of *wetting and drying* as well as *freezing and thawing* that contribute to secondary-particle formation.

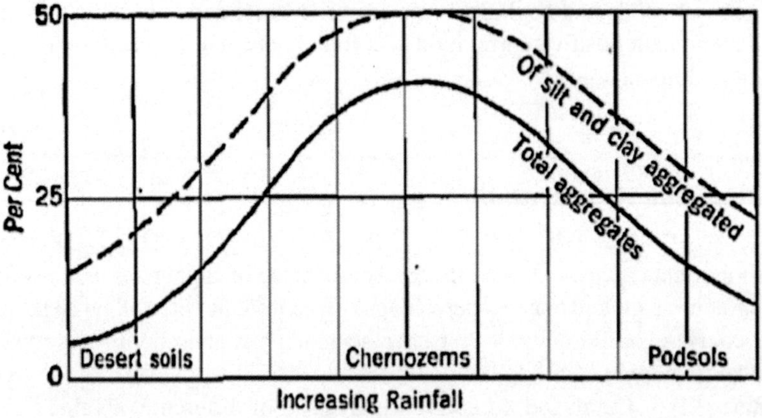

(a)

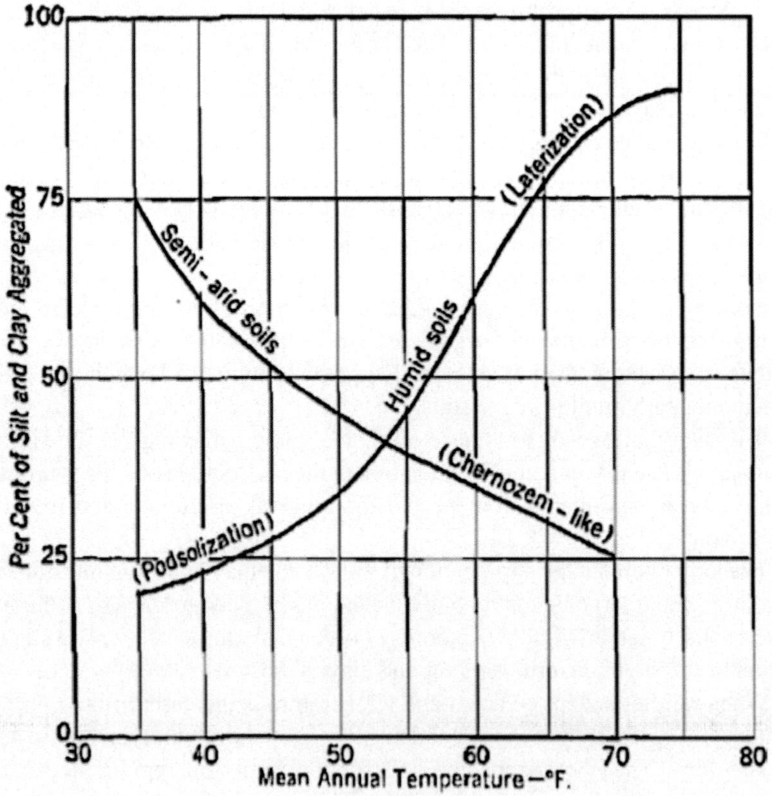

(b)

Fig. 7.4 The climatic aspect of soil aggregation: (a) temperature constant and (b) rainfall constant.

CONCLUSION

Soil structure depends on mineralogy, method of deposition, climate, topography, vegetation and human intervention. Its effect on air and water quality through storage and retention is considered very important in terms of environmental studies. Soil structure also has a major influence on the ability of soil to perform different functions. Stability or arrangement of particles are very important for functions like plant growth, groundwater recharge to building houses and roads. Soil aggregates which is the backbone of soil structure, are the primary repositories of carbon, water, microbial communities, plant nutrients, and pollutants within the soil profile.

THINK FOR A WHILE

- What is the difference between soil texture and structure?
- Is flocculation and stable aggregate formation synonymous?
- What are the cations that help in improving soil structure and how?
- What is the relationship between organic matter and clay content on soil aggregation?
- Why desert soils has low percentage of aggregate?

REFERENCES

Azuma, I., Kimura, H., Niinaka, T., Aoki, T. and Yamamura, Y. (1968), *J. Bact.*, **95:** 263–271.

Baver, L.D. (1956). Soil Physics. John Wiley & Sons, Inc., New York Chapman & Hall, limited, London.

Biswas, T.D. and Karale, R.L. (1974). Mineralogy of Soil Clays and Clay Minerals (eds. S.K. Mukherjee and T.D. Biswas), Indian Soc Soil Sci., New Delhi, India, pp. 165–179.

Biswas, T.D., Gupta, S.K. and Naskar, G.C. (1961), Water Stable Aggregates in Some Indian Soils J. Indian Soc. Soil Sci. **9:** 229–304.

Bradfield, R. (1936). The value and limitation of calcium in soil structure. Am. Soil Survey Assoc. Bull XVII: 31–32.

Daji, J.A. (1970). A textbook of Soil Science. London: Asia Publishing House.

Deshpande, T.L., Greenland, D.J., Quirk, J.P. (1964). Role of iron oxides in the bonding of soil particles. *Nature.* **201:** 107–108.

Filippovich, A.K. (1956). Absorption of colloids by soil and the formation of structure. *Pochvovedenie* No. **2:** 16–26.

Kohnknke, H. (1968). Soil Physics. New York, McGraw-Hill.

Kononova, M.M. (1966) Soil Organic Matter: Its Nature, Its Role in Soil Formation and in Soil Fertility. Pergamon Press Ltd., Oxford, 45–49.

Kroth, E. M. and Page, J.B. 1947. Soil Sci. SOC. Amer. Proc. 11: 27–34

Martin, J.P. (1971). Soil Biol. *Biochem.,* **3:** 33–41.

Martin, J.P. and Aldrich, D.G. (1955). Soil Sci. SOC. Am. Proc., **19:** 50-54.

Martin, J.P., Ervin, J.O. and Shepherd, R.A. (1966). Soil Sci. SOC. Am. Proc., **30:** 196-200.

Nikiforoff, C. C. (1941). Soil Dynamics. *The Scientific Monthly*, **52(5):** 422–429.

Peterson, J. B. (1944). Soil Sci. Soc. Am., Proc. **9:** 37–48.

Peterson, J.B. (1947). Calcium linkage, a mechanism in soil granulation. Soil Sci. Soc. Am. Proc. **12:** 29–34.

Chapter 8
Soil Tillage

"Where tillage begins other arts follow"—Daniel Webster

Tillage comes from the 'Anglo-Saxon' words *Tilian* and *Teolian*, meaning 'to plough and prepare the soil for the seed to sow, to cultivate and to raise crops'. Jethro Tull, an English agriculturist, is considered the father of tillage. He suggested that ploughing is necessary to make the soil into fine particles.

Tillage refers to different mechanical manipulations of the soil through digging, overturning and stirring that is necessary for crop production.

Tilth is the physical condition (structure) of soil achieved after tillage; it is the result of tillage. The tilth may be of three types i.e. coarse, fine and moderate.

TYPES OF TILLAGE

Different types of tillage are shown in Fig. 8.1:

Preparatory Tillage

Operations that are done to prepare the field before sowing crops are called preparatory tillage and include deep opening, loosening of the soil, incorporating or uprooting weeds and crop stubble, seedbed preparation, using tillage equipment, etc. These are costly processes but have to be used at the right time for better harvesting.

S. Mukherjee, *Current Topics in Soil Science*,
https://doi.org/10.1007/978-3-030-92669-4_8

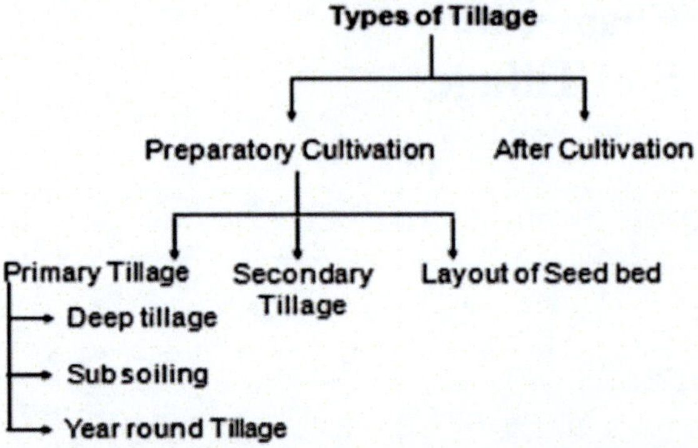

Fig. 8.1 Types of tillage.

Fig. 8.2 Comparison between deep tillage (left) and shallow tillage (right) with the help of chisel plough.

Primary Tillage

It is done mainly after the last harvesting of crops. Soils need to be wet enough to plough thus primary tillage is also done at the beginning of the next wet season. Ploughing helps in loosening and inverting the soil for better tilth. Uprooting unnecessary weeds and stubble is also a part of primary tillage.

(a) **Deep Tillage:** Tillage operations performing below the normal tillage depth to alter physical and chemical properties in the soil is called deep tillage. It is done by fracturing the subsurface compacted layer without disturbing the top soil. This helps in permitting root penetration at deeper levels. Deep tillage is only required in fields having a moisture content of less than 30–50%. It is ideal in fields under no-till conditions. Crops like pigeon pea and maize need deep tillage operations for growth (Fig. 8.2).

(b) **Subsoiling:** The process of tillage operation where compacted layers are fractured without inverting the soil. Subsoiling is done in soils below a depth of 14 inches. It helps in providing a greater volume of soil for cultivation and results in larger pores in the soil. These macro pores help in percolating excess water used in cultivation and get stored as groundwater. Breaking hardpan helps in deeper root penetration which results in a reduction in runoff and soil erosion. Chisel ploughs are used to break the hard pans without disturbing the topsoil. It is done once every four to five years.

(c) **Year-round Tillage:** Tillage operations done throughout the year is called the year round tillage. This starts with summer rainfall in dry farming regions during the preparatory stage. Repeated use of tillage operations like ploughing and harrowing is done before and even after harvesting crops. To avoid weed growth after harvesting, these operations are continued.

1.1. Secondary Tillage

After primary tillage, smaller operations are performed which are called secondary tillage. These mainly include breaking larger clods, uprooting weeds or stubble, using fertilisers or manures, etc. Harrowing and planking are used to crush the clods and smoothen the soil surface. Sowing operations are also done in secondary tillage operations.

Layout of Seedbed

After the preparation of the seedbed, the plot is laid out for sowing and irrigation. Levelling board, buck scrapers, etc. are used for levelling and markers are used for the layout. Depending on the crop type, different types of layout take place. For example for wheat, soybean, pearl millet, groundnut, castor flat levelled seedbed is prepared whereas, for vegetables and maize, the layout is done in ridges and furrows.

After Cultivation

The tillage operations that are carried out before the harvesting of crops and after the sowing or planting of crops is called after tillage/inter tillage/inter cultivation. The main reason for this tillage is to root out weeds and to reapply fertilizers and manures. This also helps in conserving the moisture levels in some soils by closing deep cracks. Earthing is also done in inter tillage to provide extra support to plants at the base by forming ridges. Spade, hoe, weeders, etc. are used for intercultivation.

MODERN CONCEPTS IN TILLAGE

Intensive Tillage Tillage operation that buries most (<15%) of its cover crop residue into the soil is called intensive tillage or conventional tillage. The crop residues are buried at an average depth of 15–30 cm. Involves multiple operations with implements such as a mould board, disk, and/or chisel plough.

Reduced Tillage Reduced tillage needs less degree and frequency of operations. The depth of tillage operations is also reduced. About 15–30% of crop residues are left on the surface. Thus it reduces soil erosion and production cost and increases soil productivity. Chisel plough, field cultivators, or other implements are involved in this type of tillage.

Conservational Tillage This type of tillage help in preserving soil and soil moisture. About 30% of crop residues are left which act as a protective cover against erosion and evaporation of soil moisture. Conservation tillage helps in improving the physical properties of soil. Fuel consumption is reduced since organic residues are used. It also controls weed and nematode populations (Fig. 8.3).

Types of conservation tillage include: No-till: The soil is left undisturbed from harvest to planting except for nutrient injection. The most powerful benefit of no-tillage is improvement in soil biological fertility, making soils more resilient.

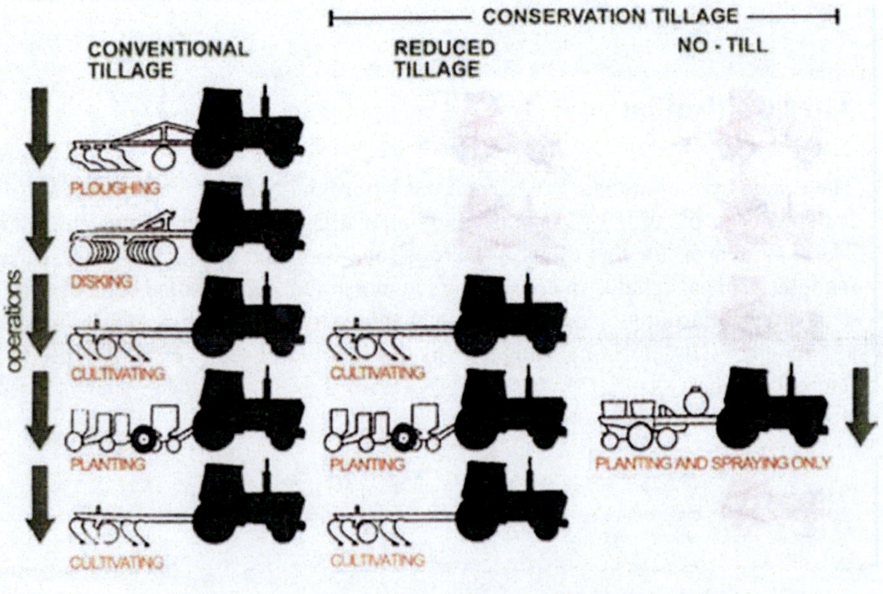

Fig. 8.3 Comparison of operations between conventional and conservation tillage.

Strip-till: It is a modification of no-till. It is used to place nutrients in bands 12.20 cm deep in the fall, to reduce soil compaction or consolidation, and/or to create a warmer and drier soil environment in spring.

Ridge tillage: The type of tillage in which planting is done on top of the ridge or both sides is called ridge tillage. Ridges are built in rows to reduce runoff and erosion. Ridges also provide better drainage and a warmer environment for the soil to dry up faster. The cost of herbicides is reduced since usage is limited only to the ridges.

Mulch tillage: The soil is disturbed prior to planting whereby crop residues are mixed with the soil and a certain amount of residue remains on the soil surface. Weed control is accomplished with herbicides and/or cultivation.

CASE STUDIES

Tilth affects all of the processes occurring in the soil matrix, the most important of which is the movement of oxygen and water needed for plant growth. It is also an important indicator of soil quality because it relates to soil structure in terms such as 'mellowness' or 'friability'. To understand the effect of tilth on crops, case studies are given in terms of its measurement and its effect on soil quality.

MEASUREMENT OF TILTH

Measurement of tilth depends on mainly two factors i.e. soil properties and method of tillage. Various scientists over the years have tested many methods to understand a way to quantify soil tilth.

Keen et al. (1930) had suggested a size distribution method of various clods where soils were screened dry in the field. Considerable importance was given to particles with sizes less than 0.625 inches in diameter.

Granulation/aggregation and porosity were given importance as measures of tilth by many Russian investigators. Soils with fine texture will have smaller aggregate than large textured soils. Porosity is more important than granulation as well aggregated soils will have a range of porosity. A soil for good tillage should have an equal amount of capillary and non-capillary pores. A soil is considered to have poor tilth if the volume of non-capillary pores is less than 10%. This is because non-capillary pores allow rapid exchange of fluids which prevents oxygen deficiency in the soil. Oxygen deficiency causes denitrification in soil which in turn results in the production of nitrous oxide (greenhouse gas). Capillary pores on the other hand help in water retention through surface tension. Sandy soils are devoid of capillary pores whereas loamy and clayey soils have lesser non-capillary pores. Long-term cultivation causes soil to get compacted and aggregation is lost.

Index of tilth is mainly obtained with the help of a penetrometer. It tests the resistance or compaction in the soil in pounds per square inch (PSI). The lower the value, the lower is the compaction in soil, the better is its soil tilth. These values are affected by many factors such as soil moisture, the presence of plant roots and the position of the compacted layer in the soil.

Studies by Shaw et al. (1942) showed that the dominant factor that affects penetrometer readings is soil moisture. Values of penetrometer increase with a decrease in soil moisture (Fig. 8.4). Also, the zone of compacted layers moves closer to the surface for soils undergoing more tillage operations.

Scott-Blair (1938) proposed that compressibility (extent of looseness of soil) be used to determine tilth. This is because tilth is better with soil having more compressibility. Methods based upon granulation and porosity offer the most promising means for characterizing soil tilth. This is because determinations of porosity include not only the total porosity but also the relative distribution of the large and small pores.

Methods of tillage were also considered a useful way to measure tilth. Studies done by Browning and Norton (1945) on Iowa soils showed that ploughing has given more yield in soil than any other method of tillage (Table 8.1). But the type of crop production also affects the yield for a particular method of tillage.

For example, it has been seen that ploughing and subsurface tillage are useful for alfalfa, sweet clover, and red clover when grown alone whereas discing was only found satisfactory on land following soybeans. Ploughing although produces a better yield but is also responsible for greater soil loss.

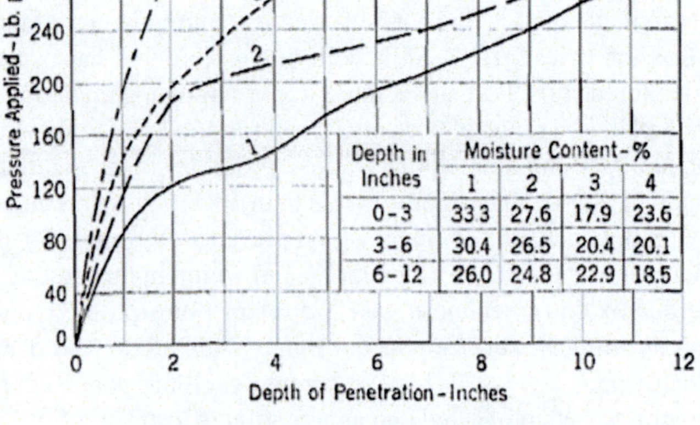

Fig. 8.4 Effect of soil moisture on penetrometer readings (Shaw et al., 1942).

Table 8.1 Effect of different tillage methods on yields of corn on different soils

	Yield of corn/acre (1944-1945)	
Tillage method	*Marshall silt loam (5 fields)*	*Marshall, Monona and Ida soils (21 fields)*
Ploughed	85.6	75.9
Hard ground listed	86.6	72.9
Subsurface tilled	82.0	70.0
Disced	75.9	68.6

Source: Adapted from Browning and Norton (1945)

Table 8.2 Effects of different methods of seedbed preparation on soil properties and yield of corn

			Physical properties of soil	
Plot number	*Method of seedbed preparation*	*8 year average yield of corn (bushels/acre)*	*Non-capillary porosity percent*	*Relative compactness*
1	Ploughed, followed by discing, standard treatment	46.7	25.9	100
2	Ploughed, complete inversion of furrow slice, corn planted directly in the surface	47.2	24.9	101
3	Rototilled	42.1	24.2	132
4	Subsurface tillage, corn planted in trash	39.1	19.2	183
5	Surface treatment only—shallow rototilling or discing only	34.5	14.2	180
6	Standard treatment, straw mulch after first cultivation	50.5	26.9	91

Source: From Page, Page et al. (1946)

Page et al. (1946) studied the effect of soil properties on different types of tillage methods (Table 8.2). It has been seen that plots undergoing subsurface and disced tillage were quite compact and poorly aerated whereas rototilled plots were loose and well aerated. Ploughed plots were less weedy than rototilled, disced and subsurface tilled plots.

Subsurface tillage was unable to kill all the sod plants which presented a major problem. Severe potash deficiency has been seen on the disced plots which have not been noticed in the ploughed plots.

EFFECT OF TILLAGE ON SOIL QUALITY

Soil Quality Indicators

Soil functions are needed to be measured to evaluate how well soil is doing. These measurements are called soil quality indicators. These indicators should be sensitive to changing climate and soil management. It should also integrate physical, biological and chemical processes in soil. Soil quality indicators should be easily accessible to all and can be easily interpreted.

Biological indicators include soil respiration, soil enzymes, earthworm activities, total C and N content in soil and particulate organic matter. Aggregate stability, available water capacity, bulk density, infiltration, slaking, soil crusts are considered physical indicators of soil quality. Chemical indicators are soil pH, nitrate, reactive carbon, electrical conductivity, etc.

Earthworms help in modifying the physical structure of the soil. It produces a binding agent that forms water-stable aggregates. The burrowing of soil by earthworms helps in the mixing of nutrients that produces new pores. This improves soil tilth, aeration, drainage and infiltration.

Other biological indicators such as soil respiration is a very useful tool to understand soil quality. It indicates microbial activity, organic content in soil and decomposition rate. Soil respiration can also infer about the nutrient cycle of the soil.

Aggregate stability is one of the major soil quality indicators. It controls other soil quality indicators. The size of aggregates indicates their function. For example, aggregates less than 0.25 mm are a more stable form of organic matter whereas large aggregates (2-5 mm) are unstable and are formed by binding smaller aggregates into larger by microbial decomposition. Thus greater amount of micro aggregates indicate better soil quality.

Karlen et al. (1994) studied the soil quality of a plot that produces corn (*Zea mays* L.) using no till method for 12 years. The same plot was then used to produce

Table 8.3 Effects of various tillage operations on some soil quality indicators of Orthic Luvisol soil samples for 12 years of continuous no-till corn production

Tillage treatment	*Wet aggregate stability*		*Total C in aggregates (g/kg)*	*Biomass (Mg C kg^{-1} soil)*	*Respiration (Mg C kg^{-1} soil)*	*Earth-worms*
	Wet sieve (%)	*Turbidity (%T)*				
No-till	45.9	1.36	24	696	352	78
Chisel	33.9	1.04	16	394	139	52
Plough	35.9	0.97	11	260	74	53

Note: Biomass: Microbial CO_2 evolved from soil fumigated with chloroform and reinoculated
Respiration: CO_2 evolved from untreated soil
log $T = \log_{10}$ of percent transmittance
Source: Based on Karlen et al. (1994)

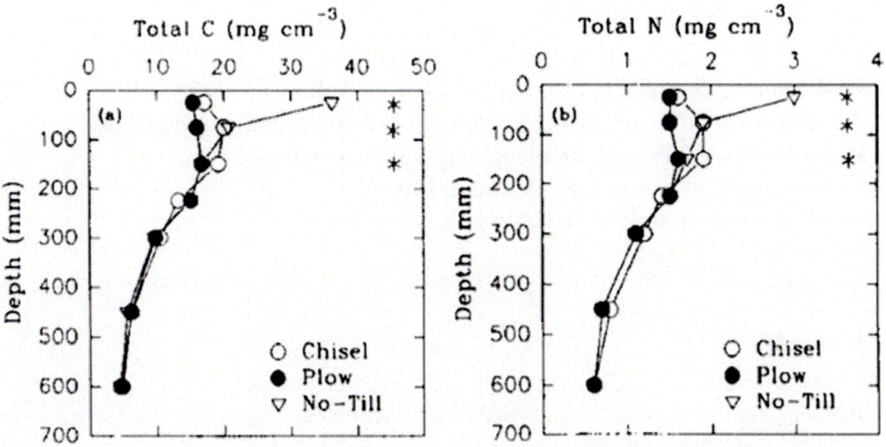

Fig. 8.5 Effect of total carbon and nitrogen content in soil with different tillage operations. The * indicates a significant difference among crop residue treatments at P~< 0.05 at the depth specified.

corn but with several tillage methods. Several physical, chemical and biological parameters of soil quality were measured. The studies showed that plots managed using no-till practices for 12 years before samples were collected for this study had surface soil aggregates that were more stable in water and had higher total carbon, microbial activity and earthworm populations than either the chisel or plough treatments (Table 8.3).

This indicated that the no till practices produced a good quality of soil compared to the same soil that had undergone various tillage operations. Figure 8.5 also suggests that total carbon and nitrogen are higher for no-till soils compared to the same soil that has undergone chiselling and ploughing.

CONCLUSION

Agriculture is the economic foundation of any country and has a major influence on components of the industry, world trade, and global ecology. Tillage of the soil has been an integral part of the agricultural production process. Thus, depending on crop type or land use different types of tillage being operated. The objectives of tilling the soil include seedbed preparation, water and soil conservation, weed control, prevention of soil erosion and improving soil structure for better plant growth. But excess tillage can cause soil erosion and moisture loss which can be reduced by decreasing the frequency and intensity of tillage (speed and depth) and reducing the size of tillage implements.

THINK FOR A WHILE

- When is it advisable to use deep tillage over shallow tillage?
- What is the difference between conventional and conservation tillage?
- Why discing is better than ploughing in some conditions?
- How does earthworm indicate soil quality?
- What are the biological indicators for soil quality?

References

Browning, G.M. and Norton, R.A. (1945). Tillage practices on selected soils in Iowa. *Soil Society of America Journal*, **10**: 461–468.

Karlen, D.L., Wollenhaupt, N.C., Erbach, D.C., Berry, E.C., Swan, J.B., Eash, N.S. and Jordahl, J.L. (1994). Long-term tillage effects on soil quality. *Soil Tillage Research*. Elsevier, **32:** 313–327.

Keen, B.A. and Staff of Soil Physics Department (1930). Studies in soil cultivation. V. Rotary cultivation. *The Journal of Agricultural Science*, **20**: 364–389.

Page, J.B., Willard C.J. and McCuen, G.W. (1946). Progress report on tillage method in preparing land for corn. Ohio Agricultural Experiment Station. Soil Science Society of America, **11**: 77–81.

Scott-Blair, G.W. (1938). A new laboratory method for measuring the effects of land amelioration processes. *The Journal of Agricultural Science*, **28:** 367–378.

Shaw, B.T., Haise, H.R. and Farnsworth, R.B. (1942). Four years experience with soil penetrometer. *Soil Science Society of America*, **7:** 48–55.

Chapter 9
Soil Water

INTRODUCTION

The hydrosphere, with a strong influence on the pedosphere, comprises the sum total of all water bodies (oceans, rivers, lakes), groundwater (renewable and fossil), and soil water. Although water is the most abundant of all resources covering 70% of the earth's surface, freshwater is a scarce resource. Table 9.1 indicates that 97.2% (volume basis) of the world water is in oceans and seas. Freshwater accounts for merely 2.8% of the total volume, of which groundwater is 0.6%, and soil water accounts for less than 0.1% of the total. Soil is a major reservoir of freshwater, which accounts for about 50 times that in rivers and streams.

PROPERTIES OF WATER AND ITS RELATION TO SOIL

Table 9.2 lists the important physical properties of water (H_2O) which are related to soil properties.

A water molecule has radius of 1.38 Å, at whose center lies oxygen atom bonding by 2 hydrogen atoms. The hydrogen and oxygen proton are about 0.97 Å apart, while hydrogen protons are 1.54 Å apart and are at an angle of about 105° from each other. This structure gives water an electric dipole of about 1.87×10^{-18} esu. Types of soil water, properties of surface tension, contact angle and capillarity are discussed as follows.

S. Mukherjee, *Current Topics in Soil Science*,
https://doi.org/10.1007/978-3-030-92669-4_9

Table 9.1 Global water resources

Reservoir/Pool	*Quantity (km^3)*	*Percent of total*
Oceans	1,370,000,000	97.2
Freshwater lakes	125,000	<0.1
Saltwater lakes and inland areas	104,000	<0.1
Rivers and streams	1,300	<0.1
Polar ice caps and glaciers	29,200,000	2.2
Soil water	67,000	<0.1
Groundwater	8,350,000	0.6
Atmosphere	13,000	<0.1

Adapted from Nace (1971); Edwards et al. (1983); Goldman and Home (1983); Van der Leeden et al. (1990); Alley et al. (2002).

Table 9.2 Properties of water

Properties of water	*Values*
Density at 20°C	998.20 kg m^{-3}
Density at 3.98°C	1000.0 kg m^{-3}
Viscosity at 0°C	1.787 centipoise
Viscosity at 20°C	1.002 centipoise
Surface tension against air at 0°C	75.6 dynes cm^{-1}
Surface tension against air at 20°C	72.75 dynes cm^{-1}
Boiling point at NTP	100°C
Freezing point at NTP	0°C
Heat of vapourisation	590 cal g^{-1}
Heat of freezing	80 cal g^{-1}

Source: Adapted from Weast (1987)

TYPES OF SOIL WATER

There are mainly three types of soil water viz Adhesion water, Cohesion water and Gravitational water (Fig. 9.1).

- **Adhesion Water:** The mutual attraction between water molecules and the surfaces of soil particles results in strong adhesive forces. If a drop of water falls onto some oven dry soil, the water molecules encountering soil particles will be strongly adsorbed and the molecules will spread themselves over the surfaces of the soil particles to form a thin film of water. This layer, or film of water, which will be several water molecules thick, is called adhesion water. Adhesion water has the lowest energy level, is the most immobile water in the soil, and is generally unavailable for use by plant roots and microorganisms.

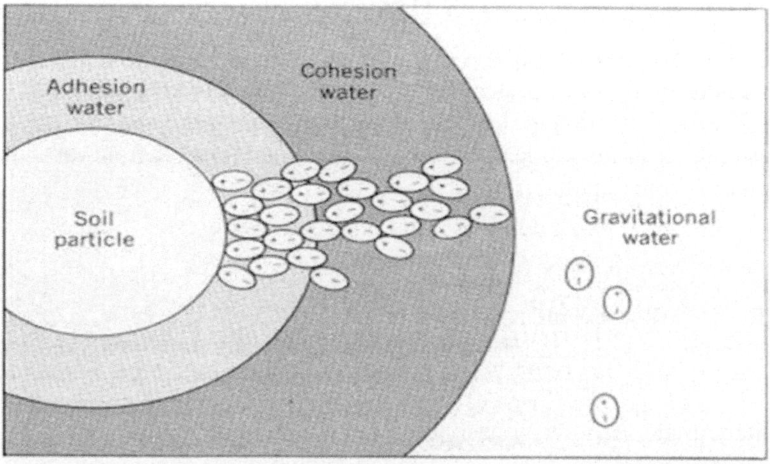

Fig. 9.1 Relationship between various forms of soil water. Adhesion water is strongly absorbed and very immobile and unavailable to plants. Gravitational water, by contrast, is beyond the sphere of cohesive forces, and gravity causes gravitational water to flow rapidly down and out of the soil (unless downward flow is inhibited). Cohesion water is intermediate in properties and is most important for plant growth.

- **Cohesion Water**: Beyond the sphere of strong attraction of soil particles for water molecules of the adhesive water layer, cohesive forces operate between water molecules because of hydrogen bonding. This causes water molecules to attach themselves to the adhesion water, resulting in an increase in the thickness of the water film. Water retained in soils because of cohesive forces is called cohesion water. Unlike adhesion, cohesion water is slightly mobile and generally available to plants.
- **Gravitational Water:** Soil water that exists in aeration pores, and is normally removed by drainage because of the force of gravity, is gravitational water. Gravitational water is not considered available to plants because it normally drains out of soils within a day or two after soil becomes very wet.

SURFACE TENSION (γ)

Surface tension is the tendency of liquid surfaces to shrink into the minimum surface area possible. The cohesive forces between the liquid molecules results in this phenomenon. The interaction of water molecules at air-water, solid-water or another fluid-water interface is subjected to different forces other than the molecules with the bulk volume of the fluid. This difference in pressure is the cause of surface tension (γ), expressed as force per unit length or dynes cm^{-1}.

THE DUPRÉ EQUATION (1969)

If a solid is immersed in water, the interfacial tension is due to the forces on adhesion (force of attraction for the water molecules by the solid and vice versa). These forces results in the work to be done to separate the solid from the liquid. The amount of work required is give by

$$W_{SW} / A_S = \gamma_{sa} + \gamma_{wa} - \gamma_{sw} \tag{9.1}$$

The work W_{SW} expressed in units of energy, and γ_{sa}, γ_{wa}, and γ_{sw} represents surface tension between solid-air, water-air and solid-water interfaces respectively, and A_S is the area of solid surface. Equation (9.1) is called the Dupré equation.

Contact angle: As already been discussed in Chapter 6, soil is a three-phase system, viz solid, liquid and gas. Suppose a fluid phase (liquid, gas) is in contact with soil solid, then the interface between air and water forms an angle called the contact angle (Fig. 9.2) This angle of contact is defined by Young's equation

$$\cos\alpha = \tag{9.2}$$

The contact angle α depends on interfacial tensions (γ).

Case 1: $\gamma_{sa} > \gamma_{sw}$, cos α is positive and $\alpha < 90°$, angle is acute and liquid meniscus is convex. This case is seen in most mineral soils and water, as water wets soils.

Case 2: $\gamma_{sa} < \gamma_{sw}$, cos α is negative and α lies between 90 and 180°, angle is obtuse and the meniscus is concave. This is the case of soil and mercury, as mercury does not wet soil.

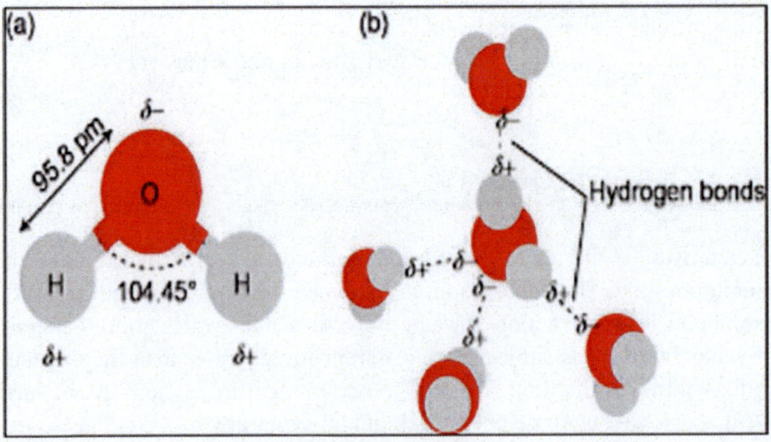

Fig. 9.2 Schematic diagram of water's atomic structure.

CAPILLARITY

$$h = \quad (9.3)$$

where h is capillary rise, γ the surface tension, r the radius of capillary tube, α the angle between meniscus and the wall of the tube, ρ is density of water and g the acceleration due to gravity (Fig. 9.3).

$$(2\gamma \cos\alpha)/r = \rho_w gh = -\Delta P \quad (9.4)$$

where ΔP is the pressure difference across the interface.

If the contact angle between solid and liquid is zero, then

$$h = \quad (9.5)$$

$$r = \quad (9.6)$$

Now, substituting values of water at 20°C in Eq. (9.6), $\gamma = 72.75$ dynes/cm or h/s^2, $\rho_w = 0.9982$ g/cm^3, g = 980 cm/s^2. Assuming α is 0, then

$$r = (0.1487/\mathrm{h})\,\mathrm{cm} \quad (9.7)$$

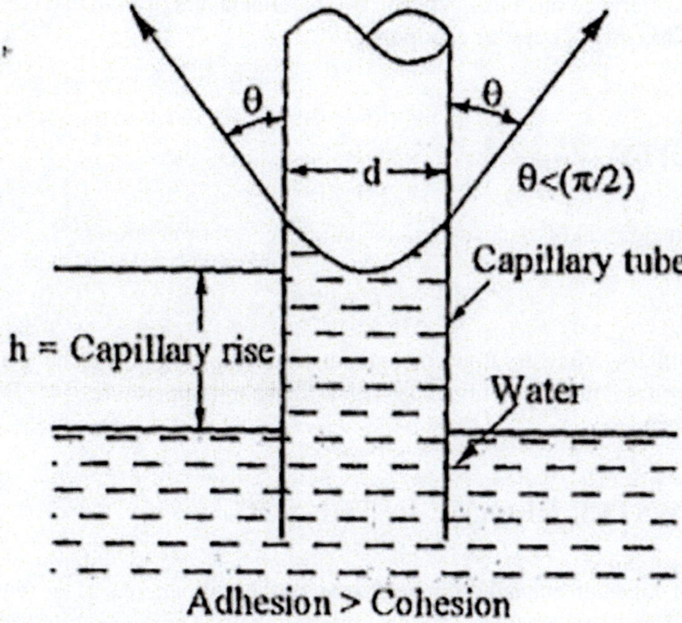

Fig. 9.3 Diagrammatic representation of capillarity.

Example 1: What is ΔP at the surface of a droplet of water and mercury with r of 1 mm?
Solution: For air-water interface, $\Delta P = (2\gamma)/r = 2(72.75/0.1) = 1455$ dynes/cm^2

VISCOSITY

The increase of soil moisture content, causes its consistency from plastic to sticky to viscous. Viscous soil flows under stress and the flow is proportional to the force applied F. The behaviour of flow is described by Bingham equation as:

$$V = k\mu(F - F') \tag{9.8}$$

where V is the volume of flow, μ the coefficient of mobility, F the force applied and F' the force required to overcome the cohesive forces. If $F' = 0$, then V is directly proportional to F and k is the constant of proportionality in viscous flow and called the coefficient of viscosity. Viscosity equation

$$\tau = F_s / \mathrm{A} = \eta(du / dx) \tag{9.9}$$

where τ is shearing stress, F_s the force, A the area of the force acting on it, (du/dx) the velocity gradient normal to the stressed area and η the viscosity coefficient. Viscosity coefficient η is defined as the force per unit area needed to maintain a velocity difference of 1 cm/s between two parallel layers of fluid that are 1 cm apart. Unit of viscosity is poise or centipoise.

FLUIDITY

It is the reciprocal of viscosity and its unit is poise^{-1} or centipoise^{-1}.

$$\mathrm{F} = \rho g / \eta \tag{9.10}$$

Fluids with low viscosity flows more readily than those fluids that have high viscosity. Water has a fluidity of about 3% per 1°C rise in temperature.

SOIL-WATER REGIME IN THE SOIL

Forces of cohesion and adhesion hold water in the soil, where surface tension, capillarity and osmotic pressure play an important role. There are three forms of soil moisture:

1. Transmission and retention pores hold the liquid water.
2. Water that is absorbed is held by cohesion and adhesion forces on the soil particles, i.e. colloidal particles (clay and organic matter).
3. Water held within the lattice structure of organic clay.

Soil-Water Content: It is also expressed as volumetric soil moisture content or soil moisture fraction, θ (g/cm^3) is given by

$$\theta = \tag{9.11}$$

where V_w is the volume of water in the soil and V_t is the total volume of soil sample.

Soil Bulk Density (ρ_b): It is the ratio of the mass of dry soil (M_d) to the total volume (V_t) of the soil,

$$\rho_b = M_d / V_t \tag{9.12}$$

The bulk density depends on the packing of particles and mineralogy. It considers both solids and pore spaces.

Particle density (ρ_p): It is the ratio of weight of a dry soil particle (M_d) per unit volume of soil particles (soil solids) (V_s) having unit g/cm^3 expressed as

$$\rho_p = M_d / V_s \tag{9.13}$$

It differs from bulk density because the volume used does not include pore spaces. Particle density and bulk density is important as it tells about the porosity of the soil.

Porosity (ε): It is the fraction of pore space occupied in the total volume of the soil

$$\varepsilon = \left(V_p / V_t\right)$$

Also,

$$V_t = V_s + V_p$$

Relation among bulk density, particle density and porosity is given as

$$1/\varepsilon = 1 + \left(\rho_b / \rho_p\right) \tag{9.14}$$

where V_p is volume of pores and V_t the total volume of soil.

Saturated water content: The maximum amount of water that can be held by soil. It is a useful method of telling how wet is the soil.

Field Moisture Capacity (FC): It is the stable point of saturation after drainage. When soil is fully saturated ($S = 0$), it is allowed to drain freely under gravity and there is no loss of water through evaporation, the soil moisture content reaches equilibrium. This equilibrium in soil's moisture content is called field moisture capacity.

The FC marks the upper limit of moisture content that the soil can hold. In this case, all macropores or transmission pores have been drained out of water and these pores have air which replaces the water. Clay content, clay minerals, porosity, texture and pore size distribution are the important factors on which FC is dependent.

Permanent Wilting Point (PWP): This point marks the lower limit of the moisture content of soil. At this point the forces of cohesion and adhesion that holds moisture exceeds the force that the plant roots exerts to draw moisture from soil. Below this point of wilting, the plant leaves wilt permanently and does not regain turgidity even when it is placed in an atmosphere with 100% humidity.

Available Water Capacity (AWC): The water that can be absorbed by plants is called available water. It varies with soil texture (Table 9.3). It is the difference in moisture content between FC and PWP.

$$\text{AWC} = \text{FC} - \text{PWP} \tag{9.15}$$

ENERGY STATE OF SOIL WATER

Like any other natural bodies, soil water possesses two forms of energy:

1. Potential energy due to its position or configuration relative to a reference point.
2. Kinetic energy due to its motion (equal to 1/2 mV^2, where m is mass and V the velocity). Also water changes its state with temperature variation which has an effect on its kinetic energy.

Thus, the total energy of soil water is defined by its equivalent potential energy, which is determined by the various forces acting on water per unit quantity.

Table 9.3 Water availability on basis of soil type

Texture	*Available water (inches per foot of soil)*
Loamy fine sand	1.10
Sandy loam	1.40
Loam	1.95
Silt loam	2.03
Clay loam	1.95

Example 2: A soil with a bulk density if 1.2 g/cm^3 has an initial gravimetric moisture content of 0.083. If its field moisture capacity is 0.25, how deep will 2 cm of rain penetrate into the soil? Assume density of water (ρ_w) is 1 g/cm^3.

Solution: Initial volumetric moisture content (θ_f) = 0.083×1.2 = 0.10

Volumetric field moisture capacity (θ_{fc}) = 0.25×1.2 = 0.30
Soil moisture deficit = (θ_{fc}) – (θ_f) = 0.30 – 0.10 = 0.20

Thus,
Depth of rain penetration = Rainfall amount/Soil moisture deficit = 2/0.20 = 10 cm

SOIL WATER POTENTIAL

Soil water potential is defined as the amount of work that must be done per unit quantity of pure water in order to transport reversibly (independent of path taken) and isothermally to the soil water at a considered point, an infinitesimal quantity of water from a reference pool (Aslyng, 1963). In other words, it is the measure of the relative potential energy of water in the soil in comparison with pure water.

Soil water is subject to the work-energy principle, which states that the work done by an object is equal to change in its energy status.

Water potential depends on:

1. Bonding of water to surface
2. Position of water in gravitational field
3. Solutes in water
4. Pressure on water (hydrostatic)

Total Soil-Water Potential

Total soil-water potential (Φ_t) is the amount of useful work per unit quantity of pure water that must be done by means of externally applied forces to transfer irreversibly and isothermally an infinitesimal amount of water from the standard state to the soil measured in units of energy, which can be expressed per unit mass, volume or weight basis.

Total soil-moisture potential (Φ_t) consists of several components given as

$$\Phi_t = \Phi_p + \Phi_m + \Phi_z +_\pi \Phi + \Phi_o \tag{9.16}$$

where Φ_p is the pressure potential (hydrostatic), Φ_m the matric potential (adsorbtion to surface), Φ_z the gravitational potential (position), Φ_π the osmotic potential (solutes) and Φ_o the overburden potential respectively.

Summary of Components of Total Soil-Water Potential

Saturated Soil

Case 1: Non swelling soil → $\Phi_t = \Phi_z + \Phi_p + \Phi\pi$
$\Phi\pi$ is zero for soils in the humid region.
Case 2: Swelling soil → $\Phi_t = \Phi_z + \Phi_p + \Phi_o$

Unsaturated Soil

Case 1: Non swelling soil → $\Phi_t = \Phi_z + \Phi_m + {}_{\pi}\Phi$
Case 2: Swelling soil → $\Phi_t = \Phi_z + \Phi_m + \Phi_o + \Phi_p + {}_{\pi}\Phi$. Φ_p is usually 0.

Tensiometers are devices that are used to measure the potential energy of soil water relative to free water in porous medium. It measures mainly matric potential (Fig. 9.4).

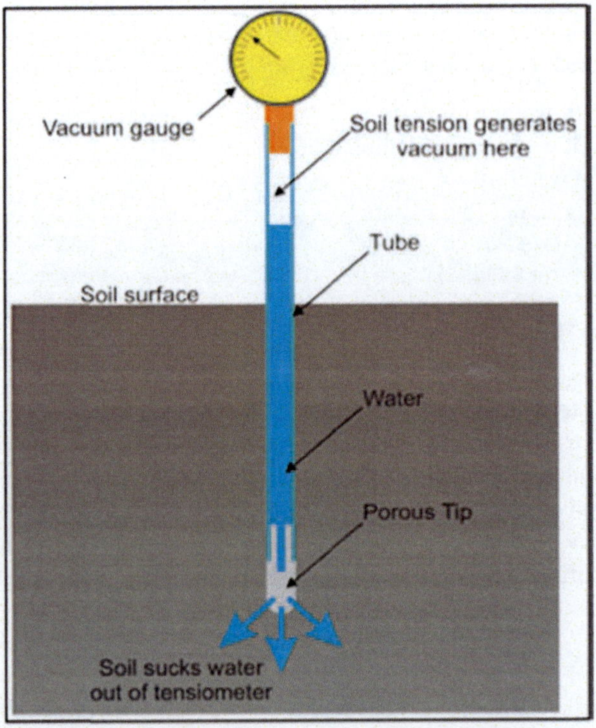

Fig. 9.4 Diagram of tensiometer.

Apparatus: It consists of a porous cup (e.g. ceramic cup) and a manometer or a pressure gauge.

Principle: The porous cup and part of manometer is filled with water and buried in soil at the desired depth. Based on the wetness of soil, the water moves from the cup into the soil and develops a negative pressure or suction that is measured by the fall in the height in water in the manometer tube or in the pressure gauge that is attached to the cup. There are various types of tensiometers, depending on the system used to measure suction created by the movement of water from porous cup to soil.

Mercury Manometer Tensiometer: It uses a combination of water and mercury (Hg).

$$\Phi_m = \left(\rho_{Hg} = 13.6 g / cm^3; \rho_w = 1.0 g / cm^3\right)$$
$$\Phi_m = -13.6 Z_{Hg} + Z \tag{9.17}$$

where Z_H is the distance from top of mercury column to the surface of the mercury in the reservoir and Z ($Z_{Hg} + Z_o$) the distance from top of the mercury column to the center of the ceramic cup.

Vacuum Gauge Tensiometer: In this Hg is replaced by a vacuum gauge. The dial is calibrated from 0 to 100, which on basis of weight range from 0 to –1000 cm (0 to –100 centibars). The useful range is about 0 to 80 kPa or 0 to 800 cm of water suction.

Example 3: If Z_o is 20 cm, Z_{Hg} is 14.2 cm, calculate Φ_m.
Solution: $\Phi_m = -12.6\ Z_{Hg} + Z_o = -12.6 \times 14.2 + 20 = -178.92 + 20 = -158.92$ cm

SOIL WATER MOVEMENT

Water movement in soil is described in terms of potential. Thus, total of soil water potential is expressed as the sum of the pressure Φ_p, gravitational Φ_z and osmotic $\Phi\pi$ potentials (osmotic potential comes into play when movement of solutes gets restricted by pores) (Table 9.4). In most cases of water flow, osmotic potential is neglected, so the total potential $\Phi_t = \Phi_p + \Phi_z$.

Water movement occurs when there is difference potential energy of water within the porous matrix. The *water potential gradient* (*h*/*l*) is the water potential difference between two points divided by the distance between the two points. The rate of water flow is directly related to the water potential difference and inversely related to the flow distance.

The velocity of water flow is also affected by the soil's ability to transmit water or the *hydraulic conductivity* (*k*). Velocity of water (*V*) flow through large pores is faster than through small pores and flow is faster when the conductivity is greater.

Table 9.4 Brief description on soil-water potential components

Components of soil-water potential	*Description*	*Unit*
Pressure potential (Φ_p) (hydrostatic)	Pressure exerted by water of the overlying saturated column of water on a specific position within a soil. $\mathbf{\Phi}_p$ = Work = pv	Work (Φ_p) per unit volume = pv/v = ρgh dynes/cm^2, where p is density (g/cm^3) and g is acceleration due to gravity (cm/s^2).
It exists only under saturated water conditions (S = 1). Positive pressure potential usually occurs below the groundwater level called the peizometric head or the submergence potential. In field condition, pressure potential is measured by a piezometer tube.		
Matric potential (Φ_m) (adsorbtion to surface)	Matric potential exists only in unsaturated soils. It is due to effects of soil solids, interfacial curvature because of cohesion and adhesion forces and surface tension of the soil matrix. This negative pressure potential is also called capillary potential. Under specific soil water condition a soil either has pressure potential (Φ_p) or matric potential (Φ_m), but not both. Matric potential is measured using tensiometers.	$\mathbf{\Phi}_m$ per unit volume = ρgh dynes/cm^2 $\mathbf{\Phi}_m$ per unit mass = gh ergs/g; $\mathbf{\Phi}_m$ per unit weight = h cm
Gravitational potential (Φ_z) (position)	It is due to the position of soil water $\mathbf{\Phi}_z = \rho hg$ (g = 9.81 ms^{-2} = 9.81 kPa	$\mathbf{\Phi}_z$ per unit volume = ρgZ dynes/cm^2; $\mathbf{\Phi}_z$ per unit mass = gZ ergs/g; $\mathbf{\Phi}_z$ per unit weight = Z cm
Osmotic potential (Φ_π) (solutes)	It is due to the presence of solutes in soil moisture that affect its thermodynamic properties (e.g., entropy, enthalphy, free energy). Presence of solutes in soil lowers the vapour pressure of soil moisture and affects its $\mathbf{\Phi}_\pi$	$_\pi\mathbf{\Phi}$ per unit volume = $\rho gZh\pi$ dynes/cm^2. $_\pi\mathbf{\Phi}$ per unit mass = $gh\pi$ ergs/g $_\pi\mathbf{\Phi}$ per unit; weight = $h\pi$ cm
Overburden potential (Φ_o)	It is due to the mechanical pressure exerted by the unsupported solid material on the soil water. The overburden pressure is usually significant only in swelling soils	

Darcy's Law

It states that, "the rate of flow of water through a porous medium was proportional to the hydraulic gradient" (Darcy, 1856).

$$Q = KA\left(dh / L\right) \tag{9.18}$$

where Q is the discharge of water, K the hydraulic conductivity, A the cross-sectional area of flow, L the length of flow or soil matrix and ($\Delta h/L$) is hydraulic gradient Δh = ($h_i - h_o$), where h_i and h_o are hydraulic head maintained at inlet and outlet of the soil matrix, respectively.

Table 9.5 Hydrological properties of soil types

Soil type	*Saturated hydraulic conductivity (cm/hr)*	*Porosity*
Sandy loam	2.59	0.45
Silt loam	0.68	0.50
Clay loam	0.23	0.46
Clay	0.06	0.475

Source: Derived from measurements of different soil types in the USA. From Rawls et al. (1982).

The relationship with volumetric flow rate through soil matrix, when soil system is homogeneous (no layering and variation with depth) and isotropic (soil properties is uniform in all direction) is given by—

$$\begin{aligned} Q &= V / t \\ Q &= KA(\Delta h / l) \\ (Q / A) &= K(\Delta h / l) \\ q &= K(\Delta h / l) \end{aligned} \tag{9.19}$$

where q is flow per unit cross-sectional area per unit time [LT^{-1}] and also called flux density or Darcy's velocity. K_s is the proportionality constant called as saturated hydraulic conductivity. When K_s become equal to q, hydraulic gradient is unity. Saturated hydraulic conductivity refers to its ability to conduct water when all pores are full with water.

Hydraulic conductivity of slow saturated soil ranges from less than 0.125 cm/hr to 0.5 cm/hr. Moderately saturated hydraulic conductivity ranges from 0.5 cm/hr to 12.5 cm/hr and rapid saturated from 12.5 cm/hr to more than 25 cm/hr (Table 9.5).

Case 1: Flow is vertical, pressure and gravitational potentials or heads vary, so the gradient of flow changes.

Case 2: Flow is horizontal, gravitational head is constant in all throughout the soil, only the pressure head is the driving force.

Water when flows through the soil pores is called the pore water velocity, which is the actual velocity of water moving through the pores and is expressed as

$$= (q / \theta) \tag{9.20}$$

where is the mean pore water velocity through soil matrix, q the flux density or Darcy's velocity and θ is volumetric water content of the soil matrix.

Validity of Darcy's Law

There are two types of flow:

Laminar flow: The state of flow where the water flows like a sheet with uniform velocity throughout ($N_{Re} < 1$).

Turbulent flow: The state of flow is in radial and axial movement ($N_{Re} > 1$).

Darcy's law is valid only when the flow is laminar ($N_{Re} < 1$, N_{Re} = Reynold's number). Reynold's number (N_{Re}) is the ratio of inertia forces to viscous forces, and is a non-dimensional quantity.

$$N_{Re} = (\text{Forces of inertia}) / (\text{Forces of viscosity})$$
$$= \rho_w rv) / (\eta_w) \left[ML^{-3}\, L\, LT^{-1} \right] / \left[ML^{-1}T^{-1} \right] \quad (9.21)$$

where ρ_w is the density [ML^{-3}] of water at a given temperature, η_w the dynamic viscosity of water [$ML^{-1}T^{-1}$], v the water velocity [LT^{-1}], r the radius of pore [L] and N_{Re} is dimensionless.

Case 1: Viscous force is high, N_{Re} remains low and flow remains laminar.
Case 2: Viscous force is low, N_{Re} becomes high and flow becomes turbulent.
Note: N_{Re} is inversely proportional to viscous force.

Instrument used for measuring hydraulic conductivity is permeameter. Constant head permeameter is used in laboratory. In field, hydraulic conductivity can be determined using peizometer and constant head permeameter.
Factors affecting water movement are: (1) soil texture, (2) structure, (3) amount of organic water, (4) depth of soil, (5) temperature and (6) pressure.

Infiltration, Percolation and Irrigation

Water Movement in Stratified Soil: Stratified soils have layers, or horizons, with different physical properties, resulting in differences in hydraulic conductivity. As a result, the downward movement of water may be altered when the downward moving water front encounters a layer with a different texture.

When soil pores are completely filled with water, saturated flow occurs. Water moves at water potential larger than –33 kPa. Saturated flow begins with infiltration where water moves into soil from rain or irrigation water. When soil profile gets wetted, the extra water moves downward through the wetted soil, which is termed as percolation.

Infiltration

Infiltration rate is measured by recording the rate at which water enters the soil. When rain falls on the soil surface, water seeps down the pores under the influence of both pressure and gravitational head potential. If rainfall intensity is less than the infiltration rate (IR) then all water is absorbed. In case of wetting dry soil, the pressure gradient is high. As the depth of wet soil increases the pressure gradient decreases according to Darcy's Law.

Wetting Front: If the soil is structurally homogeneous and the rainfall continues, the water travels downwards at a constant rate q, that determines the infiltration capacity of the soil defined by Eq. (9.20). In nature, soils are not structurally homogeneous due to presence of roots, worm holes and pore gaps between aggregates causes obstruction in smooth downward water movement.

Percolation

As infiltration continues, a considerable fraction of water percolates from surface to subsoil, forms groundwater or recharges it. The percolation effect depends on local flow conditions. In this case, hydraulic pressure difference caused by precipitation plays an important role.

When soil is a homogeneous porous system (grain size distribution), it results in formation of wetting front. When hydraulic gradient in wetting front is steeper, then it lowers the water content in that front. The pressure gradient decreases with the advancing of wetting front as the water content of soil and with it unsaturated water increases. With this the hydraulic gradient also decreases which is necessary for transport (Darcy's law).

In arid climate, there is no downward percolation occurring in the entire soil profile. Evaporation leads to upward movement of the soil solution instead. The transport caused by percolation also depends on the pore system. This relation between water content and unsaturated water conductivity dominates the gradient that is required to supply water to soil surface, thus controlling water content with depth profile.

Non-volatile components from soil water like salts, that accumulates on the soil surface is mostly restricted to the surface. Fertilizers and other agricultural substances also accumulate close to the surface. It leads to drop in water potential and results in decrease of osmotic potential.

Irrigation

When reduction of water supply occurs due to changes of the hydraulic potentials, irrigation is needed to compensate the deficiency of water. Irrigation process principle is similar to the process of infiltration attributing same relationship.

Irrigational process recharges the water capacity of soil which gets lost due to evapotranspiration. In humid climatic conditions, continuous irrigation is required for a longer period of time as water percolates to subsoil zone. The method of irrigation is used effectively in agricultural purpose. Sustainable and economic use of water for irrigation is necessary. Some of the irrigation methods are described in Table 9.6.

Water Movement in Dry Soil: Evaporation rate is controlled by soil properties and decreases with time. This stage is known as falling-rate drying. The dry surface layer has a self-mulching effect which conserves soil moisture at depth. In spite of self-mulching effect, prolonged evaporation without rain causes zone of dry soil to extend downwards.

Table 9.6 Types of irrigational methods with their advantages and disadvantages

Irrigational method	*Type of irrigation*	*Technique*	*Advantage*	*Disadvantage*
Subsoil irrigation	Capillary irrigation	Raises groundwater level	Reduces evaporation. No loss of farmland by above surface plumbing	Not suitable for shallow rooted crops. Salt accumulation as watering of topsoil.
Surface irrigation	Flooding pooling, ditch irrigation	Controlled flooding. Construction of barrages and distribution systems	Feasible in flat terrains. Unsophisticated and simple	High evaporation rates. High water requirement. High investment.
Sprinkling	Field irrigation, partial irrigation, spot irrigation	Controlled distribution of time and water by using pumps, pipes and irrigators	Independent of surface shape, multi-purpose irrigation. Saves water. Protection against freezing due to irrigation	Impaired by wind. Requires equipment and experience. High energy consumption, costly.
Drip irrigation	Point irrigation, drip irrigation	Dispersal of small amount of water around drip dispenser	Adaptable to crop. Saves water and minimizes evaporation. Lower annual recurrent costs, fertilizer supplied to crops on time	Required clear water (clogging). Technically challenging. Tillage impaired by surface or buried piping, expensive.

Flow of water through dry soil occurs by simultaneous transfer of water and dissolved salts in soil that is subjected to controlled, isothermal evaporative conditions. Slightly above PWP (permanent wilting point), chloride ions move down with water. Below PWP, water mainly moves by vapour diffusion with response to vapour pressure (VP) gradient resulted from the differences in matric and osmotic potentials.

Normally on VP, effect of osmotic and matric potential is small because the range of 0 to –1500 kPa, relative VP changes from 1.0 to 0.989. Hence, soil becomes air-dry when relative VP of soil water drops to 0.85.

Capillary Rise: Another type of water movement of water is through capillary rise. Above water table, soil water is pulled upwards through pores until the pressure gradient acting upwards is balanced by the gravitational head gradient acting downwards. This process of water rising is called capillary rise, depends on pore size and distribution of soil particles.

When solutes and salts particles accumulates on soil surface, due to mismanaged irrigated soils, it encourages capillary rise due to high osmotic pressure maintains an upward gradient even in moist soils.

CONCLUSION

The properties of water such as cohesion, adhesion, capillarity etc. governs a lot of functions in soil. Thus studying its properties in depth is very important for its application in the field of agriculture, irrigation and nutrient management. The movement of water through soil is one of the foundation processes of terrestrial life on the planet. With the increase in population with time, there is more demand of both water and food that highlights most strongly the need to sustainably manage our soil and water resources. Important conversions and equivalent quantities of 1 bar is given in Table 9.7.

THINK FOR A WHILE

For the mercury-air interface=8600 dynes/cm^2, the difference calculated across the interface does not refer to the difference due to vapour pressure, which may be substantial. What vacuum is needed to draw all the water out of a sintered glass funnel if the minimum pore size is 4 μm in diameter?

A Tensiometer dial is calibrated from 0 to 30. If the gauge is 25 inches above the tensiometer cup and it reads 20 inches of Hg, calculate Φ_m.

Why some soils are more wettable than others? Why does burning crop residue or any biomass make a soil hydrophobic?

Calculate soil moisture content of 20 g of wet sample that registers an increase in volume by 5 cm^3. Assume ρ_s of 2.7 g/cm^3.

Table 9.7 Value of 1 bar of expressing soil-moisture potential on volume basis and some important conversions

Equivalent quantity of 1 bar	*Unit conversions*
1 bar	1 Dyne = 1 g cm/s^2 = 10^{-5} N
= 100 centibars (cb, cb= 1 J/kg)	1 N = 1 kg m/s^2
	1 Pa = 1 N/m^2 (1 kPa = 1 = Jkg^1)
= 1000 millibars (mb)	1 J = 1 Nm = 10^7 ergs = watts
= 1020 cm of H_2O	1 bar = 10^5 Pa = 0.987 atmosphere = 29.53″ Hg = 10^5 dynes/cm^2
= 401.57 inches of H_2O	1 Torr = 1 mm Hg = 1/760 Atmosphere = 1013,250/760 dynes/cm^2
= 75.01 cm of Hg	= 133.22 microbars
= 0.9869 atmosphere	1 Watt = J/s = 107 erg/s
= 100 J/kg	1 erg = 1 dyne cm = g/cm s^2 = 10^{-4} J/kg
= 14.5 lbs inch^2 (PSI)	
= 10^6 erg/g	
= 10^6 dynes/cm^2 =10^2 kPa	

REFERENCES

Alley, William, Healy, Richard, LaBaugh, James and Reilly, Thomas (2002). Flow and Storage in Groundwater Systems. Science (New York, N.Y.). **296**: 1985–1990.

Aslyng, H.C. (1963). Soil Physics Terminology. Bull. *Int. Soc. Soil Sci.,* **23:** 7–10.

Darcy, H. (1856). *Les fontaines publiques de la ville de Dijon*. Paris: Dalmont.

Edwards, K.A., Classen, G.A., Schroten, E.H.J. 1983. The water resource in tropical Africa and its exploitation. ILCA Res. Rep. No. 6, Addis Ababa, Ethiopia, 103pp.

Goldman, C.R. and Home, A.J. (1983). Limnology. McGraw-Hill International Book Company, Japan, 464.

Nace, N.R. (1971). Scientific framework of the world water balance. UNESCO Tech. Papers Hydrol. 7, UNESCO Paris.

Rawls, W.J., Brakensiek, D.L., Saxtonn, K.E. (1982). Estimation of Soil Water Properties. *American Society of Agricultural and Biological Engineers.* **25:** 1316–1320.

Van der Leeden, F., Troise, F.L., Tod, D.K. (1990). The water encyclopedia. Lewis Publishers. Chelsea, MI, 808pp.

Weast, R.C., 1987. Physical constants of organic compounds. CRC Handbook of Chemistry and Physics, 68th Edition (ed.: R.C. Weast). CRC Press, Boca Raton, FL, B67–B146.

https://agwaterexchange.com/2016/08/14/understanding-the-basics-of-water-in-soil/

Chapter 10
Soil Air and Temperature

Soil air, as a part of the unsaturated zone of soil, is a necessary soil constituent for the growth of the majority of plants. Restriction of soil air can limit the development of an extensive root system. Limited soil air can also impair root respiration thereby retarding water and nutrient absorption. Moreover, essential biological activities can also be prevented hindering good fertility in the soil. All these will ultimately result in a decline in plant growth.

COMPOSITION OF SOIL

Soil air is a continuation of atmospheric air and there is a continuous exchange of gases between the atmosphere and soil air. The circulation of air in the soil mass is known as 'soil aeration'. The soil air contains a number of gases out of which oxygen, carbon dioxide, nitrogen, and water vapour are the most important (Table 10.1). The composition of soil air is more or less similar to that of the atmospheric air except for the content of carbon dioxide which is several times higher. Small deficits of oxygen and a surplus of carbon dioxide are quickly equilibrated by air interchanges between the atmosphere and soil air. This happens because a difference in atmospheric pressure is created between the soil and atmosphere for the two gases. Thus carbon dioxide diffuses out of the soil and oxygen diffuses into the soil. The diffusion rate is very rapid to prevent an oxygen deficiency or carbon dioxide toxicity for roots.

The soil fauna and flora consume oxygen and give out carbon dioxide. Organic matter addition and cropping increase the level of carbon dioxide in the soil. Carbon dioxide in soil air constantly moves from the soil pores into the atmosphere and vice-versa. The nitrogen content in soil air is almost the same as in the atmosphere.

S. Mukherjee, *Current Topics in Soil Science*,
https://doi.org/10.1007/978-3-030-92669-4_10

Table 10.1 Percentage of the volume of different gases in soil air and atmosphere

	Percentage by volume		
Types of air	*Oxygen*	*Carbon dioxide*	*Nitrogen*
Soil air	20.6	0.25	79.2
Atmosphere	20.96	0.03	79

FACTORS AFFECTING THE COMPOSITION OF SOIL AIR

1. **Soil Moisture**

Pore spaces in the soil are occupied by air and water. Thus, as the volume of soil water increases, the volume of air in soil decreases and vice versa. The oxygen content of soil decreases when the macropores are filled with water. But when the soil is artificially drained again, the macropores are filled with air and the oxygen content of the soil increases. Thus in monsoon, oxygen content in soil is lesser due to an increase in soil moisture. For well-structured soils, the air and water contents of the soil are nearly equal to field moisture capacity.

Bulk Density

It is the weight of dry soil per unit of volume typically expressed in g/cm^3. Total porosity and soil air content decrease with an increase in bulk density. Bulk density depends on several factors such as soil texture, organic matter content, soil structure and density. Bulk density is low for porous, well-aggregated soils and soils rich in organic content. This is because organic matter increases the soil water holding capacity thereby reducing bulk density. Bulk density also increases with depth since subsurface layers are more compact, less porous and have less root penetration compared to surface layers. Also, subsurface layers have lower organic matter. Land management practices like tillage increase the bulk density of soil since it destroys soil aggregation and organic matter content thereby making the soil loose. These highly erodible materials then fill pore spaces in the soil making it more compact. Sandy soil, due to its less porosity, has more bulk density than clayey and silty soils.

Microbial Activity

Soil organic matter is decomposed by microorganisms present in the soil. Microbiological decomposition leads to the production of carbon dioxide and its content increases in the soil air. Hence, soil rich in organic matter contains a higher percentage of carbon dioxide. Also, microorganisms in soil need oxygen for respiration which is taken from the soil air, thus decreasing its concentration.

Seasonal Variation

The quantity of oxygen is usually higher in the drier season since the soils are drier providing greater gaseous exchange. This causes higher O_2 and lowers CO_2 levels in the soil. Temperature also influences CO_2 levels in the soil as higher temperature increases microorganism activities in soil which results in higher production of CO_2.

Cropping

Plant roots take up O_2 from soil and release CO_2. This causes a gradual increase of CO_2 in the soil compared to O_2. Thus cropped land contains more CO_2 in the soil than fallow land. Table 10.3 shows that the land cropped with wheat contains 4 to 6 times more CO_2 in soil than fallow land. Also the use of manure increases the CO_2 in soil, which is shown in Tables 10.2 and 10.3.

Crops should be chosen by the type of aeration in the soil. Shallow rooted plants such as grasses and clovers tend to grow well in low aerated soils whereas deep-rooted plants need well-aerated soil. Many crops such as rice prefer to flourish in submerged soils.

MEASUREMENT OF SOIL-AIR COMPOSITION

Measurement of soil aeration is done in three categories i.e. capacity measurement, intensity measurement and transport rate measurement. Capacity is defined as the soil's ability to contain air. These depend on several parameters such as porosity, void ratio, etc. Capacity can be easily measured by measuring soil volume, bulk density, particle density and moisture content. This is done by wetting the soil in the

Table 10.2 Effect of cropping system on the concentration of different gases in different seasons

Cropping systems	*Seasons*	*Gases (%)*		
		O_2	*N_2*	*CO_2*
Arable land manured and cropped	Summer	20.74	79.03	0.23
	Winter	20.31	79.32	0.37
Arable land unmanured and cropped	Summer	20.82	78.99	0.19
	Winter	20.42	79.37	0.21

Source: Based on Russell and Appleyard (1915)

Table 10.3 Effect of tillage treatments on concentration of CO_2 in soil

Tillage treatment	*Percentage of CO_2 in soil air (dates)*				
	May 15	*May 25*	*June 10*	*July 7*	*July 27*
Unmanured fallow	0.10	0.07	0.08	0.08	0.09
Manured fallow	0.22	0.32	0.17	0.36	0.35
Cropped to wheat	0.61	0.32	0.35	0.48	0.30

Source: Based on Russell and Appleyard (1915)

laboratory and subjecting it to specified water tension. Later the volume of air in terms of total porosity is measured by an air pycnometer. When soil air volume falls below 10%, plant growth gets impaired.

Intensity measurement is the measure of partial pressure or concentration of certain gas in soil. It is mainly done for measuring oxygen. It is measured by withdrawing samples of soil air or analysing in situ.

Withdrawing soil air samples (0.5 ml) is done with the help of a syringe but it has a chance of contamination from the atmosphere. In situ analysis is done using several methods such as polarographic and paramagnetic instruments but they have limitations too. When O_2 concentration falls below 21%, it indicates an increase in CO_2 concentration due to the respiration of plant roots and organisms.

Transport rate measurement is performed based on the concept of diffusion and convection. Diffusion measurement is performed using the oxygen diffusion rate (ODR) method which is the rate at which O_2 is replenished in soil. It is done with the help of a platinum microelectrode. When a certain potential is applied across the platinum electrode and a reference electrode is inserted in the soil, oxygen is reduced at the platinum surface. The electric current flowing between two electrodes is proportional to the rate of oxygen reduction. Measurement of transport rate by convection is done by simple flow permeaters. It can also be measured by calculating the difference in air pressure in soil and atmosphere with the help of a hydrophobic membrane probe (HMP).

SOIL RESPIRATION AND MEASUREMENT

Soil respiration is defined as the amount of carbon dioxide flux emitted from the soil surface. Soil respiration is mainly caused due to three processes: (a) respiration of plant roots, (b) respiration of soil fauna and (c) decomposition of organic matter by soil microorganisms. It depends on many factors such as soil temperature, moisture, pH, salinity, amount of soil organic matter, etc. With increasing temperature (upto 40°C), microbial activity increases thereby increasing soil respiration. An increase in soil moisture also increases soil respiration up to a point where all pores are filled with water, thereby limiting microbial respiration. Using numerous types of respirators and respirometers, soil respiration is measured (Fig. 10.1).

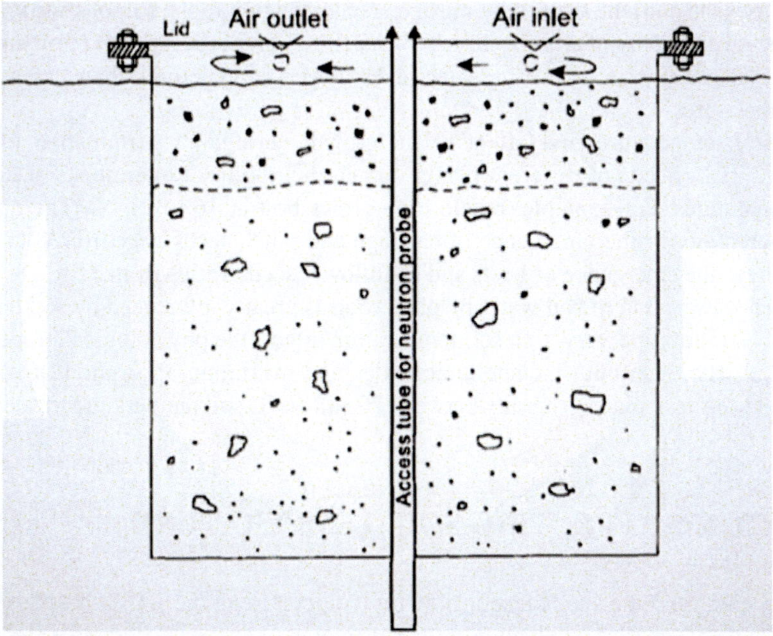

Fig. 10.1 Soil respirometer (Currie, 1975).

It is determined by the rate of CO_2 concentration increase in a chamber placed inside the soil with its opening to the soil. However, a lot of modern methods are used nowadays such as the open-flow infra-red gas analyser method (OF-method), the closed chamber method (CC-method), the dynamic closed chamber method (DC-method) and the alkali absorption method (AA-method).

The respiratory quotient (RQ) is the ratio of the volume of CO_2 produced to the volume of O_2 consumed by a soil body. For well-aerated soil, it is close to unity. It only rises to above unity when there are anaerobic pockets present in the soil. The aerobic or anaerobic conditions of soil can be checked using the following equation (Monteith et al., 1964):

$$R = R0Q^{T/10}$$

where R is the flux at T °C and R_0 at 0°C and Q is equal to 3.

SOIL TEMPERATURE

Soil temperature is a very important factor in determining plant growth. This is because, with the increase in soil temperature, the rate of decomposition of organic matter and microbial activity increases. The decomposition of the organic matter releases the nutrient, especially nitrogen contained in it in the soluble form which is either absorbed by plant roots or lost by leaching. Thus, the amount of organic matter that remains in soils is greater under lower temperatures than under higher temperatures.

Soil temperature first affects plant growth during the germination of seeds (Fig. 10.2). Seeds of different plants vary in their ability to germinate at different temperatures. For example, potato crop grows best at 16 to 21 °C. The optimum temperature for the emergence of cabbage and pinch seeds vary from 8 to 11 °C, whereas the emergence of beets and cauliflower seeds varies from 11 to 18 °C.

The absorption of soil water by plant roots is greatly influenced by soil temperature. Very low and very high soil temperature affects the physiological processes of plants. The maximum metabolic activities and maximum absorption of water by plant roots take place generally between 20 and 30 °C soil temperature, respectively.

FACTORS AFFECTING SOIL TEMPERATURE

Radiation from the sun is responsible for temperature in the soil. It depends on: (a) the angle of sun rays incident on earth, (b) amount of dust particles and water vapour in the atmosphere and (c) cloud cover. The temperature in the soil will increase with less cloud cover and when the angle of solar radiation to the earth is high. A higher amount of dust particles and water vapour in the atmosphere absorbs and reflects the heat thereby preventing it from reaching the surface of the earth.

Apart from the atmospheric condition, there are many factors on the earth's surface that affects soil temperature. They are:

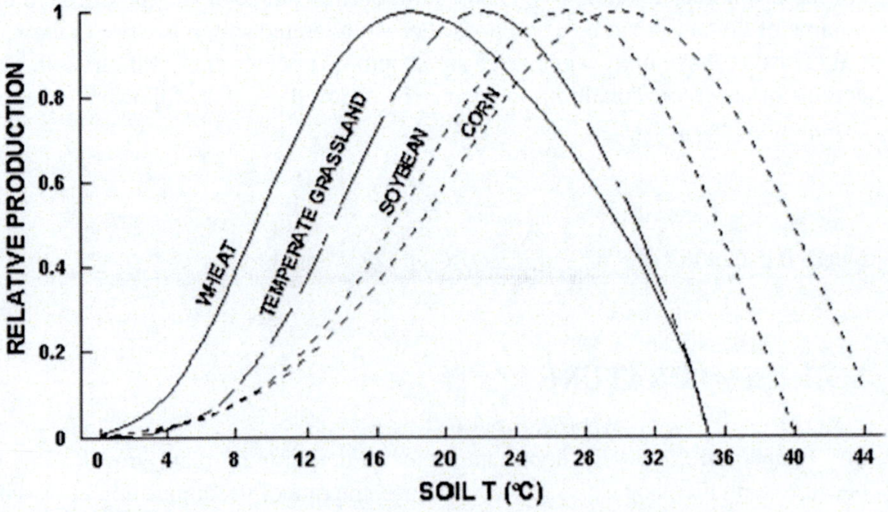

Fig. 10.2 Impact of soil temperature on potential plant growth of different crops.

Latitude The angle of incidence of sun rays depends on latitude. As latitude increases, the angle of incidence of solar radiation decreases. Heat received per unit area decreases with decreasing angle of the radiation. In other words, with slanting rays, heat is spread over a larger area compared to radiation perpendicular to the earth. Thus heat absorbed by the earth due to slanting rays is much less (Fig. 10.3). This is why temperature decreases with distance away from the equator.

Exposure and slope of land Direction of exposure influence the temperature of the soil on a larger scale. For example, in the northern hemisphere, south-facing slopes receive direct sun rays than the north-facing slopes. This causes the southern slope to absorb more heat energy per unit area (Fig. 10.4). In the north polar region, northern slopes have permanently frozen subsoil. It is not the case for south-facing slopes. The degree of the slope also influences soil temperature but on a smaller scale. The steeper the slope, the greater are the temperature differences between exposures.

Vegetative cover Vegetation cover acts as a thermal insulator and thus controls the temperature of the soil. It not only prevents solar radiation to enter the soil surface directly but also prevents losing heat from the soil to the atmosphere. Thus surface with dense vegetation remains cool during summer and warm during winter. Whereas bare soil remains heated during the day while cools off rapidly during the night (Fig 10.5). Vegetation cover also helps in losing excess heat by evapotranspiration. Other functions include altering albedo and reducing the depth of penetration of solar radiation into the soil.

Distribution of land and water Due to the high specific heat of water, it can absorb a large amount of heat. Also, the atmosphere near a water body contains a large amount of water vapour which in turn absorb solar radiation. Thus the presence of a water body can largely influence the temperature of a location. Thus continental temperatures are much higher than in coastal areas. Water currents too affect soil temperature.

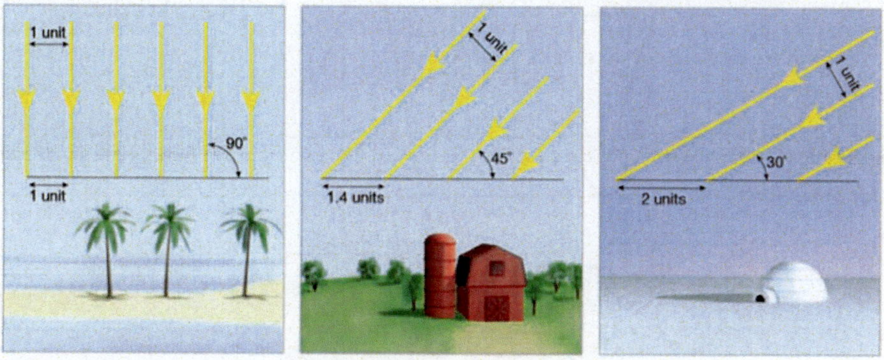

Fig. 10.3 Changes in the sun's angle cause variations in the amount of solar energy reaching the earth's surface. The higher the angle, the more intense the solar radiation.

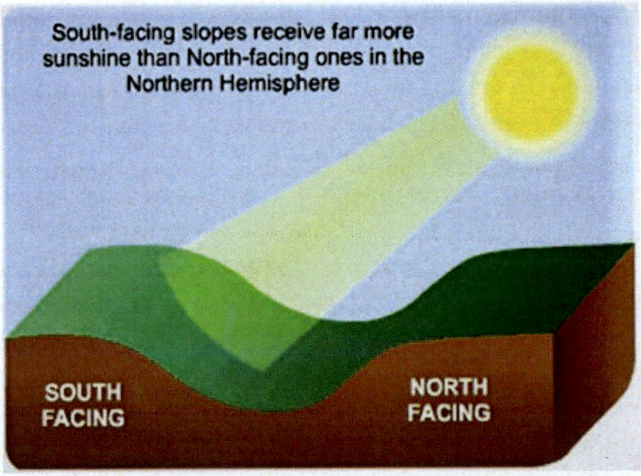

Fig. 10.4 In the northern hemisphere, soils located on south facing slopes have a higher temperature than soils on north-facing slopes.

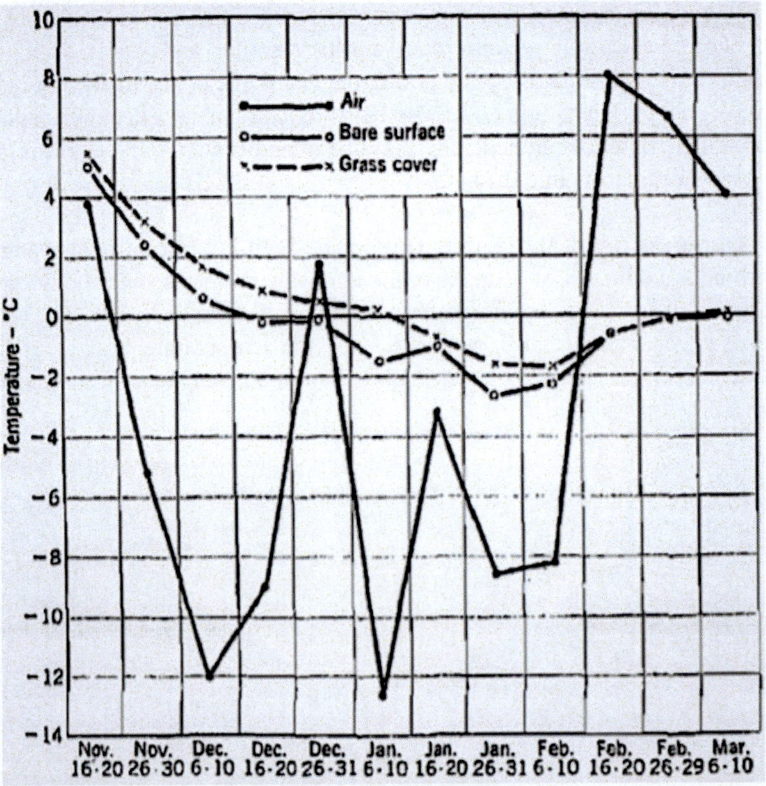

Fig. 10.5 Frost penetration in bare and grass-covered soils.

Warm water currents increase the temperature of land surrounding it while cold water currents bring down the temperature. For example, the Gulf current along the coast of Great Britain and the Japanese current along the shores of northwestern United States are warm currents that materially affect the climate of the land they touch.

BALANCE OF HEAT IN SOIL

The phenomenon of gains and losses of heat energy is called the heat balance in the soil. It is stated as (van Bavel and Hillel, 1976):

$$J_n = (J_s + J_a)(1 - \alpha) + J_{li} - J_{lo}$$

where J_n is the sum of all incoming minus outgoing radiant energy fluxes, J_s is the incoming flux of short-wave radiation directly from the sun, J_a is the short-wave diffuse radiation from the atmosphere (sky), J_{li} is the incoming long-wave radiation flux from the sky, J_{lo} the outgoing long-wave radiation emitted by the soil and α is the albedo or *reflectivity coefficient* which is the fraction of incoming short-wave radiation reflected by the soil surface rather than absorbed by it.

The albedo is an important characteristic of soil surfaces. It is defined as the degree to which a surface can reflect solar energy. It is a dimensionless quantity and varies between 0 and 1 (0 being the darkest surface and 1 being the lightest). The lighter the surface, the more energy it reflects whereas dark surfaces absorb solar energy. Albedo also depends on surface roughness, the inclination of the incident radiation relative to the surface and changing wetness of exposed soil. Albedo is higher in drier and smoother soil.

Loss of heat from soil is mainly due to the loss of radiation of solar energy back to the atmosphere. It can also happen during the evaporation of water from soil and plants. When the temperature of the atmosphere is lower than the soil, heat from soil can escape, thereby balancing the temperature difference. The heat balance is maintained for most of the time except during summer daytime where excess heat is gained. The same goes for winter nights where the loss of heat exceeds the heat gain.

MANAGEMENT OF SOIL TEMPERATURE

Management practices for controlling soil temperature depend on whether the soil temperature needs to be increased or decreased. Based on that several management practices are done such as tillage, irrigation, vegetation cover, mulching, use of dark/light substances, etc.

(A) *Light and dark coloured mulches:* Light-coloured mulches such as cotton stalk, sugarcane trash, chopped straw, plastic mulches) reduce the temperature of soil by reflecting most of the solar radiation. Similarly, dark coloured mulches such as coal increase the soil temperature by absorbing solar radiation. Dark plastic materials absorb heat but do not transmit. Thus the soil remains cool. Transparent plastic materials are used which transmits short wave radiation while reflecting long-wave radiation creating a greenhouse effect thus warming the soil. Organic mulches also increase infiltration in soil and reducing evaporation thereby maintaining the soil moisture. Soil moisture helps in keeping the soil cool.

(B) *Tillage management:* The purpose of tillage is to break the compact soil into aggregates. This causes a reduction in heat conductivity and heat loss thereby reducing soil temperature. Also, many tillage practices such as ridge tillage cause an increase in soil temperature. This is because solar radiation is concentrated in smaller areas making the soil dry. A reduction in soil moisture will increase the soil temperature.

(C) Cold, wet soils are mostly poorly drained and thus can be problematic for seed germination. This is because the soil moisture in wet soils increases the thermal conductivity in the downward direction thereby keeping the soil surface cool. Soil surfaces may also remain cool due to evaporation. Thus to increase the surface temperature, tillage practices like strip-tillage and cover crops are used. Strip tillage increases the soil temperature and excess evaporation keeps the soil dry. Cover crops on the other hand extract excess soil moisture keeping the soil dry and warm. This type of soil does not need excessive tillage operations. Studies have shown that the effect of no-tillage combined with cover crops in winter rye increases the temperature by 2.1°F.

CONCLUSION

An appropriate balance between soil air and soil water must be maintained because pores not filled with water are filled with soil air. Soil aeration influences the availability of many nutrients because microorganisms and plants that release the nutrients need soil air for survival. Soil air is influenced by moisture, temperature, bulk density, microbial activity and agricultural practices. The decomposition rate of organic matter depends mainly on soil temperature, influencing plant growth. Measurement of soil aeration, respiration and temperature thus becomes very important for the proper maintenance of soil. Management practices like mulching, ridge tillage and irrigation of excess water help to maintain healthy soil air and temperature.

THINK FOR A WHILE

- How does the composition of soil air and atmosphere differ?
- How does soil bulk density affect soil aeration?
- How is soil respiration measured?
- How does water current influence soil temperature?
- What is albedo? What are the factors that influence albedo?
- How does the colour of mulches influence soil temperature?
- How does subsurface irrigation influence crop growth?

REFERENCES

Currie, J.A. (1975). Soil respiration. *Ministry of Agriculture, Fisheries and Food. Technical Bulletin*, **29:** 461–468.

Monteith J.L., Szeicz, G. and Yukubi, K. (1964). Crop photosynthesis and the flux of carbon dioxide below the canopy. *Journal of Applied Ecology,* **6:** 321–337.

Russell, E.J. and Appleyard, A. (1915). The atmosphere of the soil, its composition and the causes of variation. *The Journal of Agricultural Science,* **7:** 1–48.

Van Bavel, C.H.M. and Hillel, D.I. (1976) Calculating potential and actual evaporation from a bare soil surface by simulation of concurrent flow of water and heat. *Agricultural Meteorology,* Elsevier, **17:** 453–476.

Chapter 11
Soil Erosion

Soil formation and soil erosion are two natural and opposing processes. Soil erosion is a natural process that can be aggravated by anthropogenic activities and is the wearing away of the earth's surface by physical forces such as wind and water. Natural agents like rainfall, flowing water, ice, wind, temperature change, gravity, etc can abrade, detach, and remove soil from one point on the earth's surface to be deposited in another location. In agriculture, soil erosion can occur due to the wearing away of topsoil through farming activities such as tillage. Soil detachment, movement and deposition are the three main actions of soil erosion.

Soil functions too can be affected due to erosion. These functions include soil buffering i.e. soil's ability to absorb and release nutrients, pollutant filtering, its role in the hydrological and nitrogen cycle, and its ability to provide habitat and support biodiversity. Out of 2000 million hectares of degraded soils in the world, 56% of the land is water-eroded, 28% is wind-eroded, whereas only 4% of the eroded land is due to physical degradation processes. The other 12% is due to manmade causes such as poor management, deforestation, etc.

Erodibility is defined as the susceptibility of soil to erosion. It is determined by the combination of physical and chemical processes in soil combined with the biological processes of the microbial soil community such as soil flora and fauna. These biochemical and biophysical processes result in the formation of aggregate which prevents soil erosion.

S. Mukherjee, *Current Topics in Soil Science*,
https://doi.org/10.1007/978-3-030-92669-4_11

WATER EROSION

Role of Soil in Hydrologic Cycle

The hydrological cycle is the continuous movement of water both above and below the earth's surface. It includes processes in which water from land and ocean moves to the atmosphere and re-enters the earth in the form of precipitation (Fig. 11.1).

Before it reaches the land surface, part of the rainfall evaporates. The water from the rainfall can get evaporated either while falling or after it has reached the land surface. After reaching the surface, the water from the rainfall either enters the soil surface or runs off or both. The type of precipitation depends on the temperature and altitude. In colder region precipitation occurs in the form of snow.

Water flowing over the surface reaches rivers, lakes and reservoirs. These waters are either stored or enters the ocean where it evaporates into the atmosphere. Water that enters into the soil can either be taken by the plants for their growth or enters into the groundwater later reaching the ocean. Apart from evaporation from water bodies, water can also enter into the atmosphere by transpiration by vegetation.

CAUSES OF WATER EROSION

Soil erosion is mainly caused due to deforestation, overgrazing, farming practices, biomass burning, shifting cultivation, mechanized logging, etc.

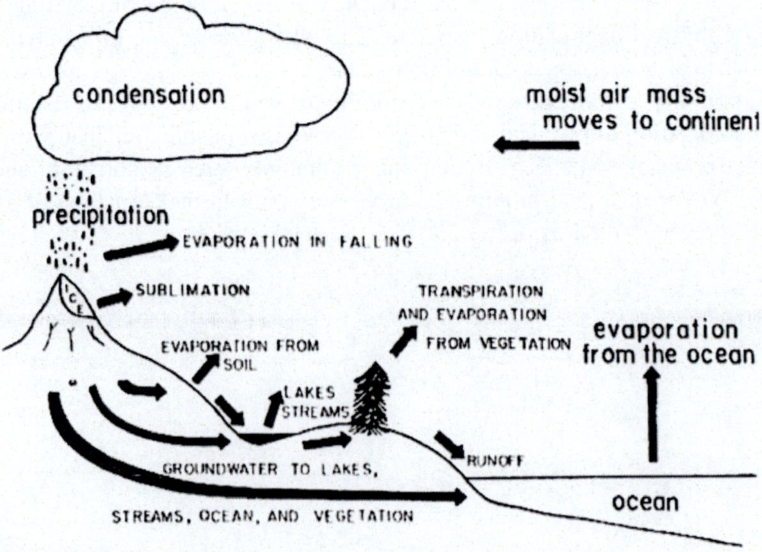

Fig. 11.1 The hydrologic cycle.

Soil erosion by water begins by mainly two processes: rainfall and runoff. These processes cause detachment or slaking in soil. Detachment is when a soil particle is physically removed and often broken into smaller pieces whereas slaking is the process of disintegration of soil particles due to absorption of water. Slaking and detachment generally occur simultaneously. The smaller soil particles then move downhill and fill soil pores and depressions. This results in decreased water infiltration rates and loss of sites where water might pond and infiltrate.

TYPES OF WATER EROSION

Water erosion is mainly of four types i.e. splash erosion, sheet erosion, rill erosion and gully erosion. Tunnel erosion is also now recognized as another common type of water erosion.

Splash Erosion: It is the first stage of the erosion process when raindrops hit bare soil. Detachment of soil aggregate takes place due to the explosive impact of raindrops into the soil. These splashed particles can rise up to 60 cm above the ground and transported to 1.5 m from the point of impact. The soil particles then block the pore spaces of the soil and form a crust resulting in a reduction in water infiltration and increase runoff. Crust formation can hinder plant growth since seedling formation and germination are inhibited (Fig. 11.2).

Sheet Erosion: It is the removal of soil in thin layers by runoff especially after the formation of crust. Fine soil particles are removed very slowly that contains most of the available nutrients and organic matter in the soil. Sheet erosion can also take place when rainfall intensity is much greater than the infiltration rate. Although the soil loss is gradual when occurs in large areas can account for large soil losses.

Little vegetation, overgrazed and cultivated soils are most vulnerable to sheet erosion.

Early signs of sheet erosion include bare surface areas, water puddling caused by rainfall, visible and exposed grass and tree roots, exposed subsoil or stony soils (Fig. 11.3).

Rill Erosion: It occurs after sheet erosion when excessive erosion causes the formation of deeper channels or rills. Rills are shallow drainage lines less than 30 cm deep. They develop due to two main conditions i.e. increase in rainfall intensity and runoff velocity. These two conditions cause an increase in erosivity thereby resulting in deeper channels. This erosion is common in hilly areas with steep slopes, bare agricultural and overgrazed land, and freshly cultivated soil where the soil structure has been loosened. The rills can usually be removed with farm machinery or can be reduced by reducing the volume and speed of surface water with grassed waterways and filter strips, ripped mulch lines, and contour drains (Fig. 11.4).

Fig. 11.2 Splash erosion.

Gully Erosion: Gullies are deeper than rills (>30 cm deep) and are mostly found in hilly areas. This is because steep slopes increase the velocity of water and cause headward/upstream erosion. Lateral erosion can also be seen in gullies which causes them to widen further.

Permanent gullies can cause a reduction in the productivity of farm plots where they are cut. Sediments eroded due to gully erosion can clog waterbodies further downstream. Thus restoration is needed to prevent further damage. These can be done by diverting the water flow in the gully to reduce scouring. It can be also done by increasing the infiltration rate and water intake by plants (Fig. 11.5).

Tunnel Erosion: Removal of subsoil by water is called tunnel erosion. This happens when water seeps into the soil through cracks and crevices and moves through the soil. This results in the dispersion of subsoil forming tunnels. With time, the tunnel becomes larger and finally collapses into a gully. This type of erosion mainly occurs in structurally unstable soils or soils high in sodium. Tunnel erosion can be very dangerous since it does not show any signs evident to the eye (Fig. 11.6).

Fig. 11.3 Sheet erosion.

Fig. 11.4 Rill erosion.

FACTORS INFLUENCING WATER EROSION

Climate: Climate is a major factor influencing erosion by water. The severity of erosion depends on its presence or absence. The climatic factors that influence run-off and erosion are precipitation, temperature and wind. Precipitation is the most important factor that causes soil erosion. Temperature affects runoff by influencing the moisture content in the soil. It also determines whether the precipitation will be in the form of rain or snow. During low temperatures, the soil freezes to an appreciable depth which causes changes in the absorptive properties of the soil. The effect of wind is mainly related to the angle at which it impacts the raindrops. It is also related to the degree of drying of soil and its consecutive need for irrigation.

Dense vegetation will be found in areas with high annual precipitation and moderate temperatures compared to sparse vegetation in low rainfall areas. These dense vegetations will help prevent soil erosion. During an annual summer monsoon period, when high-intensity thunderstorms causes flash floods, land with sparse vegetation will undergo high rates of sediment transport and eventual deposition.

Different parts of India experience different amounts of rainfall and thus the erosion profile changes. For example, coastal regions of the country experience severe rainfall and thus a high rate of erosion whereas the central part of India receives scanty rainfall thus erosion mainly occurs due to wind. So India has different erosion profiles for different parts of the country depending on the weather conditions and land management practices.

Fig. 11.5 Gully erosion.

Fig. 11.6 Tunnel erosion.

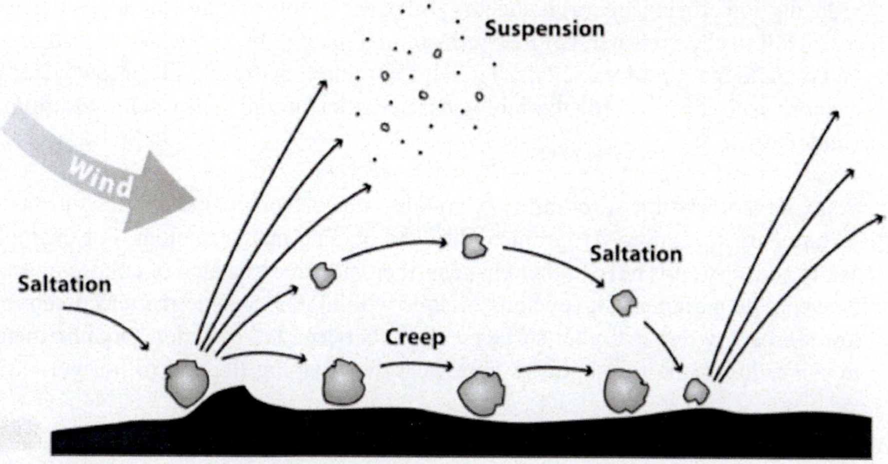

Fig. 11.7 Types of wind erosion.

Rainfall Characteristics: As discussed before rainfall is the major cause of water erosion. The size of the raindrop and its force exerted on soil particles is the measure of erosivity. Higher volume and intensity of rainfall will lead to greater soil erosion. Short-duration, high-intensity thunderstorms will lead to greater movement of soil compared to long-lasting less-intense rainfall. But less intense rainfall can lead to significant soil loss over a long period of time. Thus, the greater the intensity and duration of a rainstorm, the higher the erosion potential. Raindrop size has a positive co-relation with soil erosion. A larger drop size has higher kinetic energy and thus greater potential to disintegrate and transport soil particles.

Topography: The topographical factors that influence water erosion are the angle of slope, length of slope and shape of the slope of the land. Usually, flat lands do not have any problems with erosion. The problem starts when the topography becomes slightly rolling. The risk of erosion increases with a steeper and longer slope of the field.

This is due to the fact that the accumulation of runoff is greater on a longer slope of the field. Also, consolidation of small fields into larger ones can result in longer slope lengths. This will increase erosion potential, due to an increase in velocity of water, which permits a greater degree of scouring. In Table 11.1, it is seen that as the degree of slope increases, the rate of soil loss also increases.

The shape of the slope influences the erosion potential of water. The slope base is more susceptible to erosion than the head since runoff gains more momentum and concentration as it approaches the base of the slope. Slope shape can be concave or convex. Erosion is higher in the case of convex slope since its steepness increases towards the base. The concave slope has a flattened base and thus erosion is lower.

Vegetation: The potential for soil erosion increases if the soil has no or very little vegetative cover of plants and/or crop residues. The effects of vegetation cover on reducing soil erosion are: (a) it anchors soil particles together and increases the ability of soil to absorb water, (b) it absorbs rainfall energy by acting as an obstruction and reduces the runoff velocity, (c) it helps in removing excess subsurface water by evapotranspiration and (d) the biological activities of soil fauna helps to improve soil texture.

Soil Characteristics: Erodibility of soil depends on soil characteristics such as its texture, structure, porosity, permeability and organic matter content. For example, sandy and silty soils have higher chances of erosion due to a lack of cohesive forces between the particles. Clayey soils or soils with higher organic matter content has lower erosivity due to higher surface tension between the particles. Organic matter in soil helps in the formation of aggregate thus helping the soil to not get easily

Table 11.1 Relation of degree of slope to erosion

Soil type	*Amount of rainfall (inches)*	*Duration (hour)*	*Soil losses (tons/acre) for following slopes (%)*								
			0	*2*	*4*	*5*	*8*	*10*	*12*	*15*	*20*
Cecil clay	20	164	0.4			13.3		40.1		67.1	91.3
Putnam silt clay	3	1	2.1	3.1	6.9		11.5				
Muskingam silt loam	38.9	1934–1942 average					70.9		99.3		

Based on Diseker and Yoder (1936), Neal (1937), Borst (1945).

disintegrated. Although sandy soils are easily detachable, they have higher infiltrating capacity thus preventing runoff.

EROSION FACTORS

During the early 1900s, studies of estimation of soil erosion started. After the occurrence of the Dust Bowl in the 1930s, the US government started heavy research on soil loss and started collecting data from different stations around the country. This is to understand the factors affecting soil erosion. After 30 years of collecting and analyzing data, universal soil loss equation (USLE) was published in 1965 which was further improved. The equation is:

$$A = RKLSCP$$

where

A = soil loss (tons per acre per year)
R = rainfall erosivity factor (hundreds of foot-ton-inches per acre per hour)
K = soil erodibility factor (ton-acre-hours per hundred foot-tons per inch)
LS = slope length and gradient factor
C = cover and management factor
P = erosion control practice factor

Rainfall Erosivity Factor (*R*): The rainfall erosivity factor is the measure of the force of erosion of a specific rainfall. It is measured by the product of kinetic energy and the intensity of specific precipitation over a period of 30 minutes. These data are collected over several years and average their value. But these values are not available to many parts of the world. Thus many calculate the R factor as the function of mean annual precipitation.

Soil Erodibility Factor (*K*): The K factor measures the susceptibility of soil to erosion and the rate of runoff. It is determined by the inherent properties of soil such as texture, structure, organic matter content, permeability, etc. Soil factors that influence erodibility by water are those factors that: (1) affect infiltration rate, permeability, and total water retention capacity, and (2) allow the soil to resist dispersion, splashing, abrasion, and the transporting forces of rainfall and runoff. The *K* values for various soil types are given in Table 11.2.

Slope Length and Steepness Factor (*LS*): The slope length and steepness factor is the measure of both slope length and gradient. It describes the effect of topography on soil erosion. LS factor is the ratio of soil loss from the field slope length to that from a 72.6-foot (22.1-meter) length on the same soil type with a gradient of 9%. Slope length is defined as the point of initiation of runoff to a point where deposition occurs or where runoff enters a specific channel.

Slope steepness or gradient is defined as the angle at which the soil surface is with the horizontal datum. As the slope gradient or slope percentage increases, the velocity of runoff water also increases, which in turn increases the water's erosive power.

Cover and Management Factor (*C*): The cover management factor measures the effect of vegetation cover and land management practices on soil erosion. Land management practices include tillage, crop rotation, grazing, burning, fertilizer application, etc. It is estimated by the ratio of soil loss under actual conditions to the soil loss under reference standard conditions. Standard condition may be an area under clean tilled with continuous fallow. C factor is the most important factor among them since it is practiced by humans and thus easily managed to reduce erosion.

Erosion Control Practice Factor (*P*): The P factor is measured by all the support practices needed to control soil erosion by water. These practices include contour farming, strip cropping, terracing, subsurface drainage, compacting, etc. These help in anchoring soil particles and prevent them from being transported. It also helps in reducing the runoff velocity. It has been seen that strip cropping along with contour tillage gives utmost protection from erosion. By adopting these support practices, the P factor decreases. In other words, the lower the P factor the better the practices for reducing soil erosion.

Soil Loss Tolerance Value (*T* Value): *T* value is defined as the maximum amount of soil loss that can be tolerated without hindering plant growth. It is expressed in ton per acre per year. *T* values are small for shallow soils over hard bedrock. Thick permeable soils on permeable unconsolidated parent materials can tolerate more erosion loss. The *T* value suggests the limit of average annual soil loss that can be accepted with continuous cropping while maintaining soil productivity without requiring additional management inputs. Many soils that have developed from thick sediments of loess are agriculturally productive. But only fertilizer cannot overcome the reduction in productivity of eroded soils. These soils have a low *T* value.

Table 11.2 Erodibility and *K* values for different soil types

Soil type	*Erodibility*	*K value range*
Fine-textured; high in clay	Low	0.05–0.15
Coarse-textured; sandy	Low	0.05–0.20
Medium-textured; loams	Moderate	0.25–0.45
Medium-textured; loams	High	0.45–0.65

WIND EROSION

Wind erosion is the detachment and movement of soil particles by air moving at least 20 km per hour. It is indirectly related to water conservation in that a lack of water leaves land barren and exposed to the wind. Wind erosion is worst in arid and semi-arid areas where the following conditions frequently occur: (1) loose, dry, finely divided soil; (2) smooth soil surface devoid of vegetative cover; (3) large fields; and (4) strong winds.

Severe dust storms were seen in the United States and Canada in mid 1930s. Wind erosion along with drought destroyed the ecology and agriculture of the area.

Types of Wind Erosion

The three processes by which wind erosion takes place are saltation, suspension, and surface creep (Fig. 11.7). The major fraction of soil moved by the wind is through the process of *saltation* in which fine soil particles (0.1 to 0.5 mm in diameter) are lifted into the air by the wind and drift horizontally across the surface increasing velocity as they go. These particles are light enough to be lifted up the surface but heavy enough to be dropped to the surface again. They travel through bouncing along the surface, dislodging other particles in each impact. Saltation of soil particles can cause severe damage to the soil surface and vegetation.

Particles that are less than 0.1 mm in diameter, can be carried by strong winds easily by the process of *suspension*. These particles remain in the atmosphere and can be carried for hundreds of miles. They again deposit on the surface when the wind subsides or through precipitation.

Surface creep is a process in which larger particles (between 0.5 and 1.0 mm in diameter) are carried along the surface. These particles are too heavy to be lifted in the air. Thus they move through sliding and rolling for few meters.

Saltation carries soil particles between 50 and 75%. 3 to 40% of soil particles are carried by suspension whereas 5 to 25% are by surface creep. Thus particles moving through saltation causes most of the erosion by wind.

FACTORS AFFECTING WIND EROSION

Wind erosion depends on mainly three factors i.e. nature of wind, nature of soil and surface properties. Problems arising due to wind erosion can be solved by studying these factors.

Wind Factor: Wind erosion occurs when the wind picks up soil particles and transports them onto a different location. It can be done either by suspension i.e. moving in the air or by creep and/or saltation i.e. moving along the surface. The extent of erosion depends on wind velocity. The rate of erosion is three times higher for 30 mile/hour wind than for 20 mile/hour wind.

A minimum velocity is needed for the wind to pick up soil particles. This velocity is called the "threshold velocity". The threshold velocity of soil particles in dunes and hummocks is the lowest.

It is found that the ease at which erosion of soil occurs is dependent more on wind turbulence rather than the average velocity of the wind. The erosive power of wind is found to be a function of the rate of increase of velocity with log height and it is unaffected by surface roughness.

Chepil (1945) observed that the greatest movement of particles took place through saltation and concluded that "the movement of soil by wind is dependent, not so much on the force of the wind acting on the surface of the ground as on the velocity distribution to such height as the grains rise in saltation".

Soil Factor: The major factor that affects soil erosion by wind is the size of soil particles. The threshold velocity of wind is minimum for particles between 0.1 mm to 0.15 mm in diameter. Soil cloddiness or the ability to form aggregates is also a very important factor for wind erosion. Larger the soil aggregate, the more difficult it is for the wind to pick it up. The formation of soil clods depends on many factors. They are soil moisture content, organic matter content, clay content, compaction, etc. Soil moisture holds the individual soil particles by surface tension. Humic substances in organic matter and clay also bind the particles together to form aggregates. Thus sandy soils are easily erodible since they are most difficult to form aggregates compared to silty and clayey soils. Soil clods also act as a barrier for other erodible materials and are formed during agricultural practices like tillage.

Surface Factor: Nature of surface such as its degree of roughness and amount of vegetation cover controls wind erosion. An increase in soil roughness, decreases the wind velocity and causes deposition and trapping of saltating particles. This causes reduction in abrasion and deposition of erodible materials on the leeward side. The formation of ridges and depression causes deflection and absorption of wind energy. Ridges must be built perpendicular to the wind direction to have the maximum resistance to erosion. If widely spaced can prevent soil particles from jumping over them.

Vegetative cover is the most permanent solution for wind erosion. Vegetation helps anchor the soil particles together thus resisting erosion by wind. It also helps reducing wind velocity.

Size and type of field can affect wind erosion. Longer and broader fields with no windbreaks have higher chances of erosion. Fields with their narrower side, parallel to the wind direction and broader side, perpendicular to the wind direction have a lower rate of erosion.

CONTROL OF WIND EROSION

Principles for controlling wind erosion include:

- Stabilizing with various materials like organic crop cover residues which then anchored with a rolling disk packer.
- Producing a rough, cloddy surface by ridging and chiselling.
- Reducing field width and establish windbreaks.
- Establishing and maintaining enough vegetative cover or residue from harvested crops.
- Strip cropping i.e. the practice of farming land in narrow strips on which the crop alternates with fallow is an effective aid in controlling wind erosion.

CONCLUSION

Soil erosion has always been a major environmental threat especially in the field of agriculture. If not protected, productive soils that are easily eroded can cause crop damage. Eroded soils infused with fertilizers, insecticides, animal wastes, etc, when mixed with water, can harm the quality of that water for human use. The downstream water quality versus the relative value of productive soil determines the ecological and economic cost of erosion from land used in agriculture. Soil erosion can also send soil-laden water downstream, which can create heavy layers of sediment, eventually leading to flooding. Thus, preventing soil erosion by proper land management and agricultural practices becomes of utmost importance.

THINK FOR A WHILE

- What is slaking? What is its impact on soil structure?
- What is the rainfall factor (*R*)? Why it is a time consuming process?
- How is LS calculated for non-uniform slope?
- What is the difference between saltation, suspension and surface creep?
- How is the power of wind calculated?

REFERENCES

Borst, Harold L., McCall, A.G. and Bell, F.G. (1945). Investigations in erosion control and reclamation of eroded land at the Northwest Appalachian Conservation Experimental Station, Zanesville, Ohio, 1934–1942. U.S. Dept. Agr. Tech. Bull. 888.

Chepil, W.S. (1945). Dynamics of wind erosion: I, Nature of movement of soil by wind. Soil Sci., **60**: 305–320. II, Initiation of soil movement. *Soil Sciences*, **60**: 397–411. III, The transport capacity of the wind. *Soil Sciences*, **60**:475–480.

Diseker, E.G. and Yoder, R.E. (1936). Sheet erosion studies on Cecil clay. Alabama Agr. Exp. Sta. Bull. 245.

Neal, J.H. (1937). The effect of the degree of slope and rainfall characteristics on runoff and soil erosion. Missouri Agr. Exp. Sta. Research Bull. 280.

Part III
Soil Chemistry

Chapter 12
Concepts of Soil Chemistry

Soil is a mixture of inorganic and organic solids, air, water and microorganisms (both plant and animal in nature) (Fig. 12.1). Soil chemistry deals with the chemical composition, chemical properties, and chemical reactions of soils. This involves chemical reactions and processes between various components and particularly focuses on investigating the fate of contaminants and nutrients within soils. For example, weathering and adsorption by the soil affect air and water quality, plant roots absorb and exude or discharge organic and inorganic substances, and microorganisms catalyze many of the reactions.

Soil chemistry has always been focused on the chemical reactions in soils that affect plant growth and plant nutrition. However, in the beginning of 1970s and 1990s, concerns increased about inorganic and organic contaminants in water and soil and their impact on plant, animal, and human health. Thus, emphasis of soil chemistry went into environmental soil chemistry. Environmental soil chemistry is the study of chemical reactions between soils and environmentally important plant nutrients, radionuclides, metals, metalloids, and organic chemicals.

The fate of contaminants in the surface and soil environments can be predicted by environmental soil chemistry. An understanding of the chemistry and mineralogy of inorganic and organic soil components is necessary. This is to comprehend the array of chemical reactions that contaminants may undergo in the soil environment. These reactions, which may include equilibrium and kinetic processes such as dissolution, precipitation, polymerization, adsorption/desorption, and oxidation-reduction, affect the solubility, mobility, speciation (form), toxicity, and bioavailability of contaminants in soils and in surface waters and groundwaters. A knowledge of environmental soil chemistry is also useful in making sound and cost-effective decisions about remediation of contaminated soils (Sparks, 2003).

S. Mukherjee, *Current Topics in Soil Science*,
https://doi.org/10.1007/978-3-030-92669-4_12

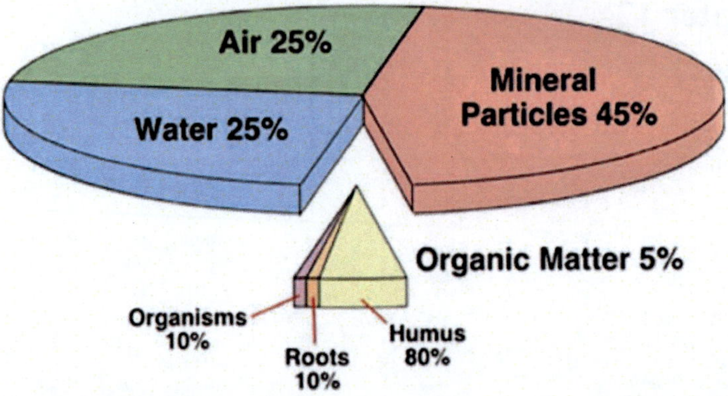

Fig. 12.1 Soil composition.

EVOLUTION OF ENVIRONMENTAL SOIL CHEMISTRY

Beginning of soil chemistry originated with J. Thomas Way's research on ability of soil to exchange ions in the early 1850s. Soon his studies spread onto soil acidity/alkanity, retention of ions by soils, clay minerology and chemistry of oxides and hydrous oxides and laid the groundwork for future researchers between 1850 and 1900. During the early decades of the 20th century classic ion exchange studies by Gedroiz in Russia, Hissink in Holland, and Kelley and Vanselow in California extended the pioneering investigations and conclusions of Way. Since the 1970s studies on rates and mechanisms of heavy metal, oxyanion, radionuclide, pesticide, and other organic chemical interactions with soils and soil components; the effect of mobile colloids on the transport of pollutants; the environmental chemistry of aluminum in soils, particularly acid rain effects on soil chemical processes; oxidation-reduction phenomena involving soils and inorganic and organic contaminants; and chemical interactions of sludges (biosolids), manures, and industrial by-products and coproducts with soils have been prevalent research topics in environmental soil chemistry (Sparks, 2003).

Evolution of soil chemistry from traditional chemical reactions affecting plant growth to a focus on soil contaminant reactions took numerous environmental movements. The modern environmental movement began over 50 years ago when the emphasis was on reducing pollution from smokestacks and sewer pipes. Focus on toxic compounds started a second movement in late 1970s which led to enactment of numerous environmental laws in United States such as the Clean Air Act of 1970, the Clean Water Act of 1972, the Endangered Species Act, the Superfund Law of 1980 for remediating contaminated toxic waste sites, and the amended Resource Conservation and Recovery Act (RCRA) of 1984, which deals with the disposal of toxic wastes. Naturally, farmers, businesses and homeowners questioned the regulations since the measures that were taken to ensure the regulations were expensive. This started a environmental wave as the critics charged that the laws were too strict and businesses and industries had to go through exorbitant cost burdens to rectify the pollution problems.

Despite the problems on the inappropriateness of laws, the public remained concerned about the quality of environment and would do anything to make the environment clean and safe.

BASIC ELEMENTS

The basic elements in chemistry is composed of atoms, elements, molecules, ions and compounds (Fig. 12.2). Elements are the basic unit of matter whereas atoms are the smallest part of an element that cannot be broken down by chemical means. Atoms combine to form molecules and a collection of identical molecules that consist of two or more different kinds of elements is called a compound. An ion is an atom or molecule that has a net electrical charge (Fig. 12.3).

Chemical Reactions

Chemical reactions occur when atoms (reactants) are rearranged to form new molecules or compounds (products). There are mainly four types of chemical reactions (Table 12.1).

Redox reactions are paired oxidation and reduction reactions. They are very common and important in nature. Oxidation occurs when an element or molecule loses an electron, and reduction occurs when another element or molecule gains the electron. The electron donor is said to be oxidized and the electron acceptor is reduced. A common redox reaction occurs in the soil when ammonia is added:

$$NH_3 + 2O_2 \rightarrow HNO_3 + H_2O$$

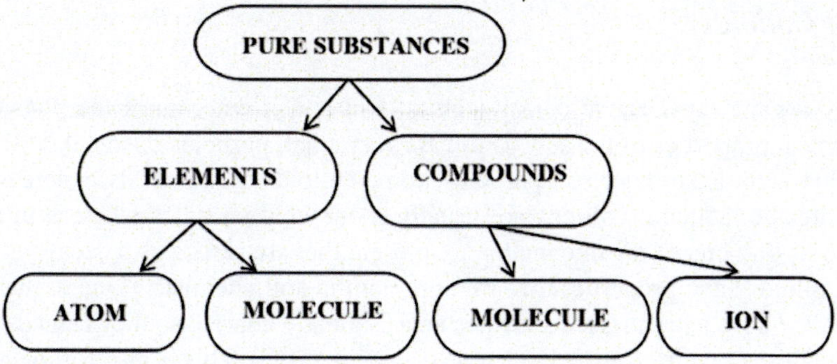

Fig. 12.2 Basic elements in chemistry.

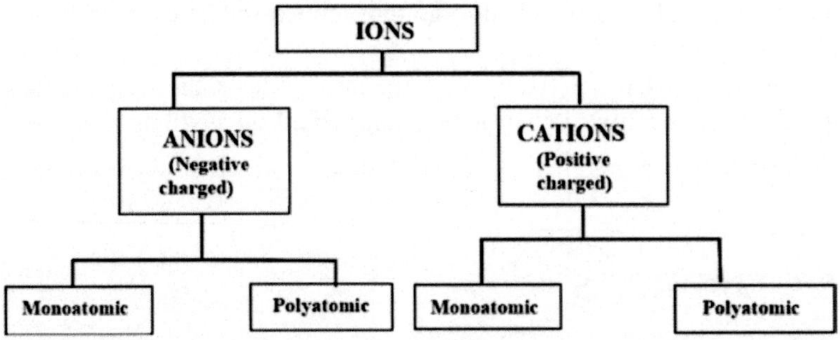

Fig. 12.3 Classification of ions.

Table 12.1 Types of chemical rections

Name	*Chemical equations*
Combination or synthesis	$S+O_2 \rightarrow SO_2$
Decomposition	$CaCO_3 \rightarrow CaO+CO_2$
Substitution or Single Replacement	$Zn+2HCl \rightarrow H_2+ZnCl_2$
Metathesis or Double Displacement	$HCl+NaOH \rightarrow NaCl+HOH$

In the presence of oxygen, ammonia (NH_3) is oxidized to form nitric acid (HNO_3).

Absorption vs Adsorption: Adsorption and absorption are two similar soil science terms with almost opposite meanings.

Absorption: is a physical or chemical phenomenon or a process in which atoms, molecules or ions enter some bulk phase—liquid or solid material.

Adsorption: is the condition in which the substances like gas, liquids or dissolved solids loosely adhere or stick to the surface of another substance which can be solid or liquid.

Absorption is an endothermic process (absorbs energy) whereas adsorption is exothermic (releases energy).

Soil Colloids

They are the most active constituents of soil and it determines the physical and chemical properties of the soil. These are very small particles of size 0.0001 mm or 0.0004 inch. Like other soil particles, some colloids are minerals, whereas others are organic. Mineral colloids are usually refine clay particles whereas in organic colloids tiny bits of organic matter are present that are resistant to decay.

Colloids have properties that are important in soil chemistry. Due to their small size they have a high relative surface area that has a charge, so they can adsorb cations. This is key for Cation Exchange Capacity (CEC). This is also for maintaining the structure of the soil (binding particles together) and for its water-holding capacity (higher concentration of colloids means greater ability to hold water).

Soil Solution

Soil solution is the liquid phase into which some of all solids and gases are dissolved. The dissolution and precipitation of solid phases largely determine the soil composition. While plants tend to get their nutrients from the soil solution, the solution does not contain sufficient nutrients at any one time to last the life of plant. Usually these nutrients are replenished from the pool of exchangeable nutrients (those that are adsorbed onto colloids). Still more nutrients are held in what is called the stable pool (bound up in solid form as minerals or organic matter).

Soil pH

The pH stands for "potential of hydrogen" and it is expressed as the negative of the log of the concentration of hydrogen (H^+) ions. Soil pH is a measure of the acidity or basicity (alkalinity) of a soil. In acidic soils (pH < 7), the H^+ ions predominate while in alkaline soils (pH > 7), the OH^- ions predominate. Soils with pH of 7 are neutral (pure water) (Table 12.2).

Acidic soils usually occur where rainfall leaches the cations out of the soil over time. Soils in arid and semi-arid areas can lack enough rainfall to leach cations, especially calcium (Ca^{2+}), magnesium (Mg^{2+}), potassium (K^+) and sodium (Na^+), from the soil. These cations bind many of the CEC sites, blocking hydrogen (H^+) ions from binding and making the soil alkaline. This can also happen if irrigating with water high in calcium bicarbonate or magnesium bicarbonate.

Soil pH affects plant nutrient availability by controlling chemical forms of different nutrients and influencing the chemical reactions they undergo. The optimum pH range for most plants is between 5.5 and 7.5. Soil microbes have reduced activity in

Table 12.2 Soil reactions and pH

Reaction	*pH*	*Reaction*	*pH*
Ultra acid	< 3.5	Neutral	6.6–7.3
Extremely acid	3.5–4.4	Slightly alkaline	7.4–7.8
Very strongly acid	4.5–5.0	Moderately alkaline	7.9–8.4
Strongly acid	5.1–5.5	Strongly alkaline	8.5–9.0
Moderately acid	5.6–6.0	Very strongly alkaline	> 9.0
Slightly acid	6.1–6.5		

Source: The United States Department of Agriculture, Natural Resources Conservation Service.

low pH soils. This can cause them to take much longer to release necessary nutrients, such as N, P, and S, from organic matter (Fig. 12.2).

CONCLUSIONS

Basic concepts of soil chemistry such as the atomic structure and bonding although preassumed to be known is briefly mentioned to bring back to memory for the readers. Evolution of soil chemistry should be known to understand the present advancement of this field. Topics like ion exchange, pH etc. is mentioned briefly so that it can easily be studied in detail in the next chapters.

THINK FOR A WHILE

- What is environmental soil chemistry? Why is it so important?
- How did soil chemistry evolve through time?
- What are the different types of colloids?
- How does soil pH affect plant nutrients?

REFERENCE

Spark, D.L. (2003). Environmental Soil Chemistry, Academic Press.

Chapter 13
Chemical Composition of Soil

Chemical composition of soil decides the chemical behaviour. Chemical behaviour is defined as the totality of (physico-) chemical reactions occurring between these constituents (and possibly other materials added to the soil *in situ*) (Bolt and Bruggenwert, 1978). The soil components are given in Fig. 13.1.

Chemical reaction type decides the reaction rate. Reactions in soil vary from a half-time of minutes for e.g. certain adsorption reactions to one of centuries for others. The very slow reactions such as the soil formation itself has already been discussed. This chapter discusses chemical behaviour of soil to the relatively fast reactions (e.g. approximately half of the growing season) that will be responsible for the composition of the soil *in situ*.

SOLID PHASE COMPONENTS

Inorganic Components

The inorganic components are listed in Table 13.1 on the basis of primary and secondary minerals in soil and its solubility. Solubility of a solute is the property of that solute to dissolve in a solvent at particular temperature. When a solute dissolves, the action takes place only at the surface of each particle. Breaking a solute into smaller pieces increases its surface area and its rate of solution. Thus, solubility increases with decrease of surface area.

Considering soil as a biotope, it is divided into three classes, viz. 'high' solubility covering those salts that in a saturated solution will inhibit plant growth because of the high osmotic pressure; 'intermediate' solubility, not exhibiting osmotic inhibition of plant growth but the solubility being high enough to play a significant role in determining the ionic composition of the soil solution; and the remaining class of 'low' solubility.

S. Mukherjee, *Current Topics in Soil Science*,
https://doi.org/10.1007/978-3-030-92669-4_13

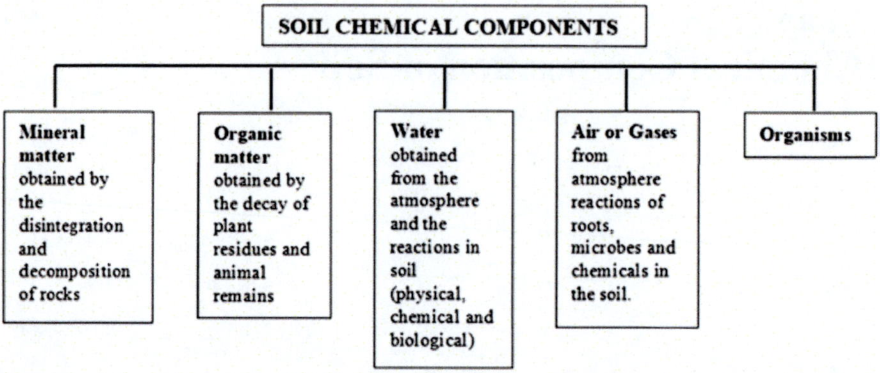

Fig. 13.1 Chemical components of soil.

Table 13.1 Common primary and secondary minerals and their solubility

Primary minerals	*Chemical formula*	*Solubility*
Quartz	SiO_2	Low
Muscovite	$KAl_2(AlSi_3O_{10})(F,OH)_2$	Low
Biotite	$K(Mg,Fe)_3(AlSi_3O_{10})(F,OH)_2$	Low
Felspar	$KAlSi_3O_8\text{-}NaAlSi_3O_8\text{-}CaAl_2Si_2O_8$	Low
Amphibole	$NaCa_2(Mg,Fe,Al)_5(Al,Si)_8O_{22}(OH)_2$	Low
Pyroxene	$(NaCa)(Mg,Fe,Al)(Al,Si)_2O_6$	Low
Olivine	$(Fe,Mg)_2SiO_4$	Moderate
Rutile	TiO_2	Low
Zircon	$ZrSiO_4$	Low
Tourmaline	$(Ca,Na,K,[vacancy])\ (Li,Mg,Fe^{+2}, Fe^{+3}, Mn^{+2}, Al, Cr^{+3}, V^{+3})_3$ $(Mg,Al, Fe^{+3},V^{+3}, Cr^{+3})_6\ ((Si,Al,B)_6O_{18})\ (BO_3)_3\ (OH,O)_3\ (OH,F,O)$	Low
Secondary minerals	*Chemical formula*	*Solubility*
Clay minerals	$Si_4Al_4O_{10}(OH)_8$ (Kaolinite)	Low
Gypsum	$(CaSO_4 \cdot 2H_2O)$	Moderate
Jarosite	$KFe^{3+}{}_3(OH)_6(SO_4)_2$	Low
Calcite	$CaCO_3$	Moderate
Dolomite	$CaMg(CO_3)_2$	High
Goethite	$FeO(OH)$	Moderate
Haematite	Fe_2O_3	Moderate
Pyrolusite	MnO_2	Moderate

High solubility components are the nitrates, most halides and some sulfates. These are present in sizable amount in mainly saline soils. Intermediate/moderate solubility are those minerals that changes its solubility with pH conditions such as some range of salts, the most common being gypsum and several carbonates such as calcite, which dissolves as bicarbonate at favourable conditions with respect to pH and CO_2 pressure and ferrous sulfate under reduced conditions. Rest of the components such as different kinds of silicates and phosphates are considered low solubility. They are considered left overs of the parent materials, i.e. minerals derived from rock formations. These residual minerals play a very minor roles in the chemical behaviour of soil. They are important in slowly delivering to the solution phase certain elements necessary for plant nutrition, usually at a very low to extremely low concentration level (Table 13.1).

Primary minerals are those which, since their crystallization, from molten magma, have undergone little or no change of chemical composition. Primary minerals are of igneous or metamorphic origin or to a lesser extent sedimentary.

Most common primary minerals in soil are the feldspers. They are found in sand and silt fractions in soils and also in clay fractions. Feldspars are anhydrous three-dimensional aluminosilicates of linked SiO_4 and AlO_4 tetrahedra that contain cavities that can hold Ca^{2+}, Na^+, K^+, or Ba^{2+} to maintain electroneutrality (Huang, 1989). Feldspars can be divided into two main groups: alkali feldspars (ranging in composition from $KAlSi_3O_8$ to $NaAlSi_3O_8$) and the plagioclases (ranging from $NaAlSi_3O_8$ to $CaAl_2Si_2O_8$). Olivines, pyroxenes, and amphiboles are known as accessory minerals in soils and are found in the heavy specific gravity fractions. Pyroxenes and amphiboles are ferromagnesian minerals with single and double-chain structures, respectively, of linked silica tetrahedra (Huang, 1989). Olivines are green neosilicates in which Mg^{2+} and Fe^{2+} are octahedrally coordinated by O atoms (Fig. 13.2). They are prevalent in igneous rocks, are sources of soil micronutrients, and are generally present in quantities smaller than those of pyroxenes and amphiboles (Huang, 1989).

Secondary minerals are formed by the disintegration of primary minerals through physical, chemical, and biological weathering. They are mainly found in the clay and fine-silt fractions because the particle size of primary minerals usually decreases during weathering. Due to small size of clay particles, they have large specific surface area and since they are charged, they have large specific charge density too. These properties contribute to the important processes in soil such as ion exchange, nutrient availability and fixation, soil physical properties such as soil structure, hydraulic conductivity, infiltration and soil management.

Aluminum, iron, and manganese oxides play extremely important roles in the chemistry of soils. While they may not be found in large quantities, they have significant effects on many soil chemical processes such as sorption (absorption and adsorption) and redox because of their high specific surface areas and reactivity. The general term "oxides" refers to metal hydroxides, oxyhydroxides, and hydrous oxides (Sparks, 2003).

Other secondary minerals include the carbonate and sulfate minerals that are highly soluble compared to silica soil minerals and are most prevalent in arid and semiarid regions (Doner and Lynn, 1989).

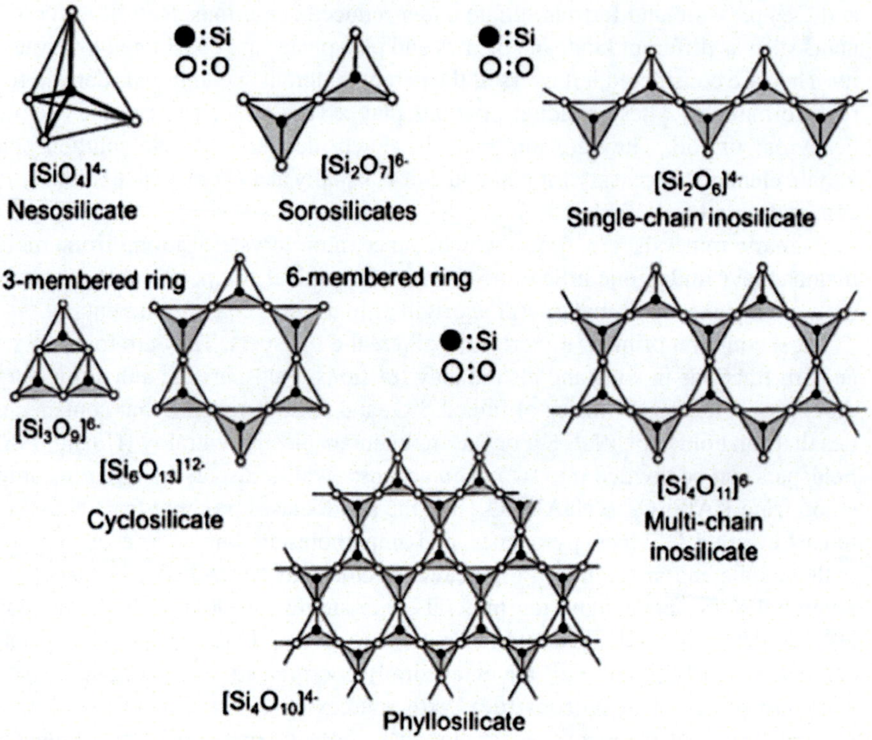

Fig. 13.2 A variety of silicates composed of $[SiO4]^{4-}$ as a fundamental structural unit and the various binding modes of $[SiO4]^{4-}$.

Organic Components

Soil organic matter (SOM) consists of a mixture of plant and animal residues in various stages of decomposition, substances synthesized microbiologically, and/or chemically, from the breakdown products, and the bodies of live and dead microorganisms and their decomposing remains (Schnitzer, 1999).

Only the non-living components are considered as the organic components of soil, i.e. the chemical and biological decay mainly of plant materials. The end product of decay of plant materials is called *humus* and it depends on the climate and local anthropogenic practices. From a physico-chemical point of view, humus may be considered as an amorphous compound with varying external dimensions (coiling and stretching of the polymer chains), a large internal surface area (in the coil), a variable charge and a strong tendency to form complexes with certain cations (Bolt and Bruggenwert, 1978). The C/N ratio decreases in humification process from excess of 20 for 'fresh' material to a value of 8–15 for its end product. No definite chemical formula exists for humic substances instead they are defined in terms of solubility behaviour in aqueous solution at different pH ranges.

Humic substances are organic compounds that are important components of humus. It consists of mainly three fractions: *humic acids, fulvic acids*, and *humin*. Soil humic acids are generally soluble in alkali and insoluble in acidic solution but a large number of organic compounds insoluble in acidic solutions are not soil humic acids or substances. Fulvic acid is soluble in both acid and alkali solution whereas humin which is insoluble in both. They makes up about 20% of humic substances.

For crops, humic and fulvic acids are chelators. They combine minerals to make them into organic compounds that can be ingested by plants more easily and also enable the soil to hold more water and can increase the water infiltration of the soil. The lower molecular weight gives fulvic acid an excellent ability to cross the membranes of plant cells so that it can easily penetrate into leaves. This makes fulvic acid particularly efficient when used as a foliar spray to provide nutrients to plants quickly.

THE LIQUID PHASE

The liquid phase of the soil is an usually dilute, aqueous solution of common salts from the ions Na, K, Mg, Ca, C1, NO_3, SO_4, HCO_3, etc. In may also contain small amounts of many other ions, depending on the presence in the solid phase of compounds with low solubility. Also organic compounds will be present by the decay of the soil organic matter. As some of these organic compounds form complexes for example with heavy metal ions, the organic constituents of the soil solution may contribute significantly to the mobility of such ions in soil. Finally, all types of organic and inorganic 'pollutants' may be present in the soil solution as a result of human activities in, on or near the soil.

The composition of soil solution is impossible to specify since soil water being highly mobile as part of the hydrological cycle, i.e. additions of water at the soil surface (rain, irrigation water, waste water) and disappearance from the soil via plants and the vapor phase, and as drainage water joining the groundwater or surface water. For detailed discussion of properties of soil water see chapter 9.

THE GAS PHASE

As discussed in chapter 10, the pore volume consists of the sum of soil moisture and air. The composition of the soil gas phase strives continually towards that of the earth's atmosphere, i.e. 78% N_2, 21% O_2, and 1% rare gases, via diffusion processes. The second process governing the composition of the soil gas phase is the consumption of O_2, and production of CO_2, by living organisms. The two processes together lead towards a roughly stationary state, corresponding to a gradual increase of the CO_2, concentration with depth from about 0.03% at the soil surface to 1–5%

underneath the plant rooting zone. The increase of CO_2 is then accompanied by a decrease of O_2, with depth (Bolt and Bruggenwert, 1978). Finally, it is seen that several organic pesticides exhibit fairly high vapour pressures. Thus, occasionally noticeable amounts of these compounds appear in the soil gas phase.

CASE STUDY

The chemical composition of soil effects soil properties. Thus, soil type changes with change in physico-chemical behaviour. Studies conducted by Dölarslan et al., aimed to determine the relationship between soil properties and plant diversity in semiarid grassland of Eldivan, Turkey. Soil properties such as sand-silt-clay contents, pH, bulk density, electrical conductivity, $CaCO_3$ and soil organic matter have been studied. It showed that different parent materials which are responsible for different soil types played an important role in determining the diversity and distribution areas of plant communities. Sand-clay-silt content, $CaCO_3$, soil organic matter content and pH of the soil changed depending on the parent material. The highest SOM content and nitrogen concentration was in the parent material of the red chalk with 34.35 g/kg whereas the lowest in serpentine with 7.71 g/kg. This is the reason that the soil with parent material of red chalk have more plant diversity and richness of plant species than soil with serpentine parent materials.

CONCLUSIONS

The variation of natural soil type and quality decides the use of the field. But, if needed, the soil type can be altered using controlled artificial methods (a rice field may be converted to an airport when needed). The interactions of climate, topography, and biota operating over geologic time on a parent rock contributes to the chemical composition of soil. The chemical components of a soil is based on its solid, liquid and gas phase. Solid phase components consists of inorganic and organic components. Inorganic components comprises primary and secondary minerals and organic components consists of mixtures of animal and plant residues. The moist gas phase consists of liquid phase, i.e. soil moisture and air.

THINK FOR A WHILE

- What is the relation between chemical reaction and rate?
- How does soil divide in terms of solubility? Give an example of a mineral for each kind.
- Why small sizes of clay is so important in many processes of soil science?

- How are humic substances divided in terms of different pH ranges?
- What are the problems that generate in terms of soil solution?

REFERENCES

Bolt, G.H. and Bruggenwert, M.G.M. (1978). Soil Chemistry, Volume 5A: A. Basic Elements (Developments in Soil Science). Elsevier Publishers.

Doner, H.E. and Lynn, W.C. (1989). Carbonate, Halide, Sulfate, and Sulfide Minerals. Ch. 6, p. 331–378. *In:* J.B. Dixon and S.B. Weed (ed.), Minerals in Soil Environments, 2nd Edition.

Huang, P.M. (1989). Feldspars, olivines, pyroxenes, and amphiboles. *In:* J.B. Dixon and S.B. Weed (Editors), Minerals in Soil Environments. Soil Sci. Soc. Am. Book Series 1, Madison, WI, pp. 975–1050.

Schnitzer, M. (1999). A Lifetime Perspective on the Chemistry of Soil Organic Matter. Advances in Agronomy. Academic Press. **68:** 1–58.

Spark, D.L. (2003). Environmental Soil Chemistry. Academic Press.

Chapter 14
Ion Exchange

Ion exchange in soil is a reversible chemical reaction which involves interchange of equivalent amount of ions between two phases in equilibrium in contact. Such exchange takes place between a soil solid phase and soil solution phase. When cations are involved, such a process is called *cation exchange* and *anion exchange* when anions are involved.

The first scientific study of ion exchange was by two English agricultural chemists J.T. Way and H.S. Thompson in 1850, who independently observed the replacement of calcium in soils by ammonium ions. Initially this phenomenon was called *base exchange* but on recognition that hydrogen ions can also be exchanged, it was widely termed cation exchange (Fig. 14.1).

CATION EXCHANGE

Cation exchange is the interchange between a cation in solution and another cation on the surface of any negatively charged material, such as clay colloid or organic colloid (Foth, 1990). *Humus* and *clay* mainly contain the macroions responsible for the cation exchange.

- Humus contains proton-donating functional groups on the surface of its molecules which dissociates as pH rises to form a negatively charged surface.
- Clay consists of fine platy crystals of aluminosilicate minerals whose lattice has a positive-charge deficiency caused by the isomorphous substitution of Al^{3+} for Si^{4+}, or Mg^{2+} for Al^{3+} in the structure of the mineral, which gives the crystal a net negative charge.

These negatively charged surfaces, also called *micellar surfaces*, form a boundary where negative charges are gathered. Thus, cations are concentrated along the boundary to neutralize the negative charge. Concentration of cations decreases as the distance from the boundary increases since the negative charge is strongest near

S. Mukherjee, *Current Topics in Soil Science*,
https://doi.org/10.1007/978-3-030-92669-4_14

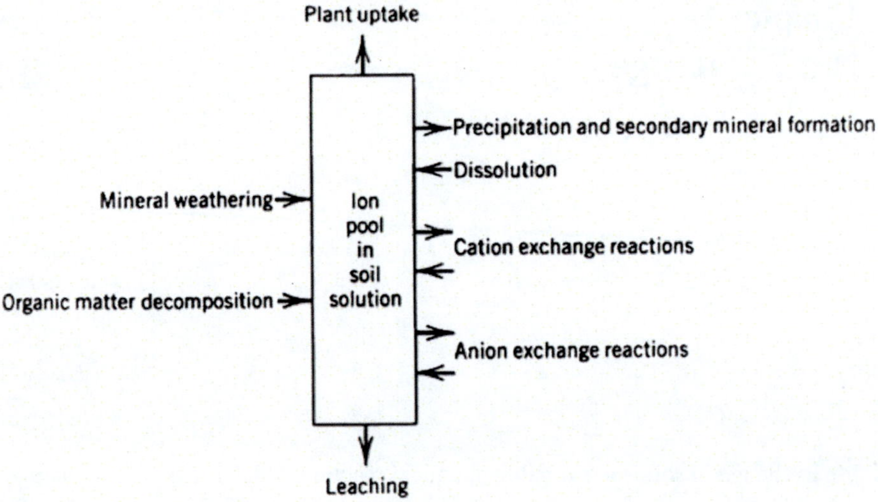

Fig. 14.1 Ions are supplied to the ion pool by weathering of minerals and decomposition of organic matter whereas ions are lost from the pool by plant uptake and leaching. Some ions precipitate and some form secondary minerals, especially clay minerals, and are subject to dissolution. Ions also absorb/adsorb onto and desorb from cation and anion exchange sites.

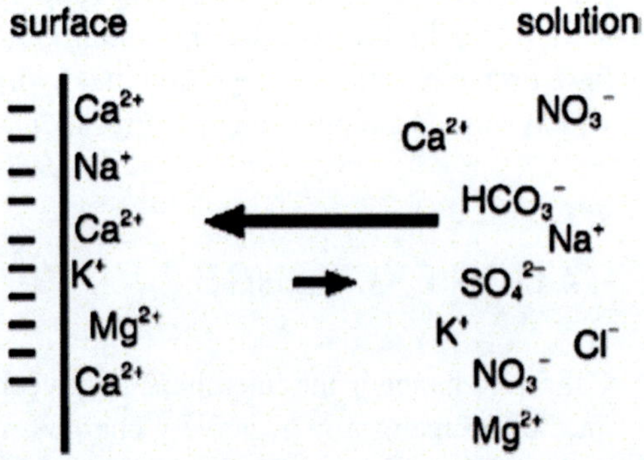

Fig. 14.2 The equilibrium between the surface and the solution.

the boundary. Conversely, concentration of anions increases with distance away from the micellar boundary. The exchange or adsorption can take place by chemically equivalent basis. For example, one K^+ replaces one Na^+, and two K^+ are required to replace or exchange for one Ca^{2+}.

Equilibrium is established rapidly between the number of cations adsorbed and the number of cations in solution although the number of cations in solution is much smaller than the number adsorbed since roots adsorb cations from soil solution and upset the equilibrium (Fig. 14.2). Exceptions are in saline soil where the content of soluble salt is much higher. Roots tend to excrete H^+ ions that increases the acidity of the soil. These H^+ ions then get adsorbed by the micellar surface. The H^+ is weakly adsorbed onto clay (ionic bonding) but it is strongly adsorbed to carboxyl groups of humus (covalent bonding).

Relation Between Cation Exchange Capacity of Soils (CEC) and pH

Cation exchange capacity (CEC) is the measure of total amount of cations that can be retained on soil particle surfaces. The sum of the positive charge on these cations equals the negative charge on the soil surface. Its SI unit is millimoles of charge per kg of dry soil, i.e., mmol $(+)kg^{-1}$ for the cations and mmol$(-)kg^{-1}$ for the surface charge. The CEC values in mmol kg^{-1} may vary for clay minerals from 30 for kaolinite to 1200 for smectite, and for soils from 50 for oxisols to 2000 for histosols (Bache, 2006).

As discussed before, the CEC of soils is affected by the amount and type of clay and the amount and degree of decomposition of the organic matter but the most important variable affecting CEC is the pH of the soil since the negative charge on humus is strongly pH dependent. CEC is directly related to pH, therefore, acid soils have a CEC less than the maximum potential CEC. Changes of CEC with pH is important for highly weathered or acid soils as CEC is much lesser in those soils. Thus, the pH of measurement should always be stated for CEC data. Conventionally, while measuring CEC, pH is considered neutral (pH 7) with an ammonium acetate buffer and ammonium NH^{4+} as the displacing cation, or at pH 8.2 with a triethanolamine buffer and barium Ba^{2+} as the displacing cation.

Exchangeable Cations

The negatively charged surfaces of humus and clay particles hold the cations by electrostatic force. Plant nutrients exist as these cations (e.g. magnesium, potassium, calcium). Thus, soils with large quantities of negative charge are more fertile because they retain more cations (McKenzie *et al.,* 2004), however, productive crops and pastures can be grown on low CEC soils.

The dominant exchangeable cations in most agricultural soils are calcium (Ca^{2+}), with lesser amounts of magnesium (Mg^{2+}), sodium (Na^{+}), and potassium (K^{+}) (Table 14.1). The acid cation aluminium (Al^{3+}) is dominant in acid mineral soils (Lanna soil) whereas hydrogen ion (H_3O^{+}) is dominant in acid organic soils.

The energy of adsorption of specific ions is related to *valence* and *degree of hydration*.

- The energy of adsorption of divalent cations is about twice that of monovalent cations.
- For cations of equal valence, the cation with the smallest hydrated radius is most strongly adsorbed because it can move closer to the site of charge (Foth, 1990). Similarly, water molecules too are more strongly attracted to smaller ions because they can move closer to the charge that is located at the center of a small ion (Table 14.2).

Table 14.1 The CEC values and major exchangeable cations of selected soils

Soils	*pH*	*CEC (mmol kg^{-1})*	*Exchangeable cation (% in total)*					
			Ca^{+2}	*Mg^{+2}*	*K^{+}*	*Na^{+}*	*H^{+}*	*Al^{+3}*
Average of agricultural soils (The Netherlands)	7	383	79.0	13.0	2.0	6.0		-
Average of agricultural soils (California)	7	203	65.6	26.3	5.5	2.6		-
Chernozem or Mollisol (Russia)	7	561	84.3	11.0	1.6	3.0		-
Sodic Merced soil (California)	10	189	0.0	0.0	5.0	95.0		0.0
Lanna soil unlimed (Sweden)	4.6	173	48.0	15.7	1.8	0.9		33.6
Lanna soil limed (Sweden)	5.9	200	69.6	11.1	1.5	0.5		17.3

Source: F.E. Bear, ed. (1964). Chemistry in the Soil, 2nd ed. American Chemical Society, Washington, DC, p. 167.

Table 14.2 Increasing order of few properties of cations

Properties	*Increasing order*
Unhydrated diameter of monovalent cations	Cs<Rb<NH_4<K<Na<Li
Unhydrated diameter of divalent cations	Mg<Ca<Ba
Adsorption strength of monovalent cations	Li<Na<K=NH_4<Rb<Cs
Potassium fixing strength of cations	H<NH_4<Ca<Mg<Na
Fixed potassium release strength of cations	Na<Mg<Ca<NH_4

Thus calcium is adsorbed more strongly than sodium because of its divalency and smaller hydrated radius (Fig. 14.3). The energy of adsorption sequence found in soil is Ca >Mg > K > Na. As a result of the small energy of adsorption of sodium, it is more likely to exist in the soil solution and can be removed from the soil by leaching whereas calcium is common as exchangeable cation.

IMPORTANCE OF CATION EXCHANGE

Plant Nutrition

Large amount of plant nutrition exists as cations such as the base cations (Ca^{+2}, Mg^{+2}, K^{+}, Na^{+}), ammonium NH^{+4}, the heavy metal cations like iron (Fe^{+2}), copper (Cu^{+2}), zinc (Zn^{+2}) and manganese (Mn^{+2}). Thus, low CEC soil often suffer from deficiency of nutrients and are susceptible to leaching than high CEC soil. Plants take up their nutrients as simple ions from solution, but when the solution is

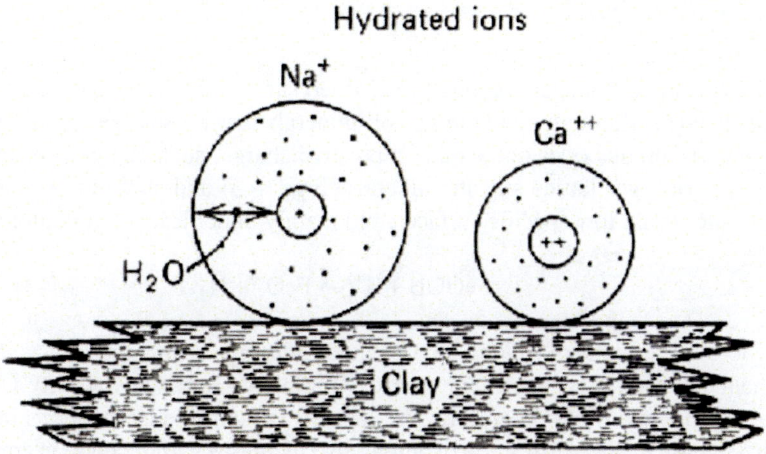

Fig. 14.3 Calcium ions are more strongly adsorbed by clay than sodium because of its divalency and smaller hydrated radius.

depleted, *exchange desorption* allows the adsorbed cations to provide a reserve from which the solution may be replenished. To maintain electrical neutrality, desorption of one ion must be accompanied by adsorption of another (Bache, 2006). Higher the CEC, the more cations can be supplied. This is called soil's *buffer capacity*.

Soil Fertility

As discussed before, soil pH is positively related with CEC. Thus, lower CEC soil which becomes acidic and leached, needs to be limed to reach the optimum pH. The CEC can also affect the frequency of nitrogen and potassium fertilizer applications. Fall applications of ammonium and potassium on low CEC soil could result in some leaching below the root zone, particularly in the case of sandy soils with low-CEC subsoils. Thus, spring fertilizer application may mean improved production efficiency. Also, multi-year potash applications are not recommended on low-CEC soils.

Amelioration of Saline and Alkali Soils

Excess of sodium salts and/or exchangeable sodium in these soils causes problems. Thus, gypsum ($CaSO_4$) is added prior to leaching the soil with excess water. Calcium from the $CaSO_4$ dissolves and exchanges for sodium (higher charge, smaller size), which is then leached in the drainage water.

ANION EXCHANGE

Anion exchange can occur on positive charged surfaces in the same manner as cation exchange. Although there are no soil minerals that possess permanent positive charges and humus does not acquire a positive charge, such charge sites exists on the edges of layer lattice silicate minerals (Fig. 14.3) and surfaces of oxides and hydroxides (with low specific surface area) mainly under low pH conditions:

$$FeOOH + H^+ \rightarrow FeO^+ + H_2O$$

The total amount of anions held exchangeably by a unit mass of soil (anion exchange capacity) is much less than CEC. Anion Exchange Capacity (AEC) is inversely related to pH. Therefore, the capacity of soil to adsorb anions increases with decrease in pH. Adsorption of anions also increases with increase in concentration of anions. Soils with dominant kaolinite and hydrated iron and aluminum oxides (tropical-subtropical soil) exhibit higher AEC. Selectivity of positive sites with respect to different anions is much complicated than the negative sites for cations. On the basis of experimental data, the preferences of anions for positively charged sites follow the order:

$$SiO_4^{4-} > PO_4^{3-} > SO_4^{2-} > NO_3^{-} = Cl^{-}$$

The AEC allows the anions (chloride Cl^- and nitrate NO_3^-) to be resistant to leaching from the soil. This provides a reserve for plant nutrition. Phosphate mainly bonds to both neutral and positively charged sites by hydroxyl ligand exchange rather than by diffuse-ion swarming; hence it is not genuinely considered "exchangeable".

ELECTRICAL DOUBLE LAYER

A electrical double layer is a structure that appears on the surface of an object (solid, liquid, gaseous object) when it is exposed to a fluid. Due to chemical reactions, a surface of an object may be charged. To balance this, a second layer of counter-charged ions is formed from the fluid.

The term was first developed by *Helmholtz* in his model where he considered the double layer consisting of two oppositely charged layers at a fixed distance apart. Thus, the electrified interface consists of two sheets of charges, one on the surface of the object (rigidly held layer) and the other in the solution (layer of oppositely charged ions). The electrical double layer leads to difference of potential between solid and liquid and that variation is linearly dissipated with distance from the surface into the solution. Helmholtz model is not satisfactory in explaining all the features since it assumes a fixed layer of oppositely charged ions in solution, but, in reality, such a rigid array of charges at an interface may not exist in solution due to thermal motion of liquid molecules.

Guoy-Chapmann's model suggested that opposite charged ions are not rigidly held, but tend to diffuse into the liquid phase until the counter potential set up by their departure restricts this tendency. The distribution of +ve and –ve charges in the diffused region is not uniform due to thermal agitation and the free movement of ions present in the solution. The potential change from one layer to another is not linear but exponential. This model assumes that ions behave as point charges and that there is no physical limit for the ions in their approach to the surface, which is not true.

Stern's model suggested that ions have finite size and thus cannot completely approach the surface of the object, i.e. the ions will be a radius away from the surface. Stern merged the above two models and stated that the double layer has two layers within, i.e. one is fixed on the surface and the other oppositely charged layer consists of two parts. One part is approximately of single counter ion thickness which remains almost fixed and compact to the solid surface (Helmholtz model). The remaining compensating charges are distributed through the fluid next to this layer in the form of diffused and movable atmosphere (Guoy's model) (Fig. 14.4). Since there are three layers in the Stern's model, there has to be two potential drop too. The second potential drop between the stern layer and the outer diffused layer is called *zeta potential*.

The main application of electrical double layer is the zeta potential which measures the effectiveness of the surface charge and the main characteristics of the

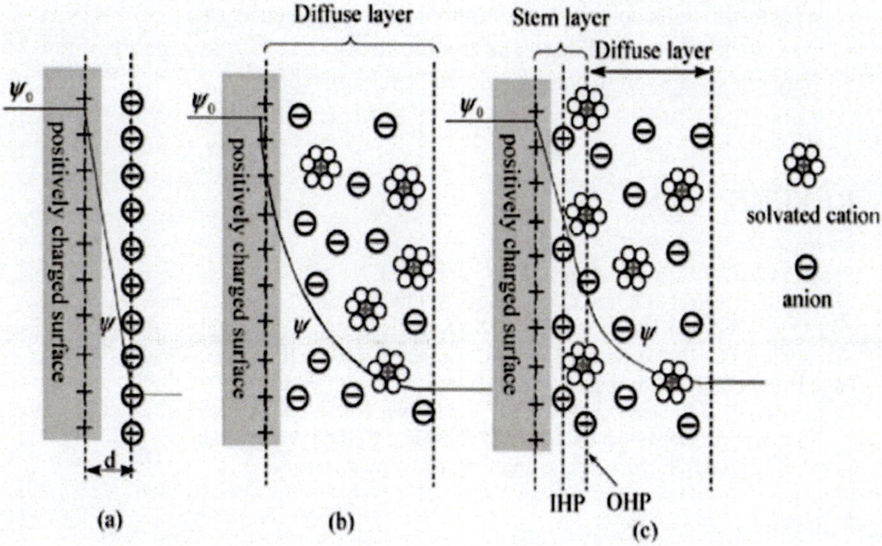

Fig. 14.4 Models for the electrical double layer at a positively charged surface: (a) Helmholtz model, (b) Gouy-Chapman model, and (c) Stern model.

electrostatic repulsion preventing particle aggregation. High zeta potential value means stronger repulsion and more stable colloidal system.

CONCLUSIONS

The CEC of soil depends on amount and type of clay, amount and degree of decomposition of the organic matter and most importantly the pH. The adsorption of specific ions is also related to valence and degree of hydration. In general, plants absorb as many anions as cations but availability to plants of the anions like nitrate, phosphate and sulfate is related to mineralization from organic matter, as well as anion exchange. Clay minerals and the sesquioxides (oxide containing three atoms of oxygen with two atoms of another element)/hydroxides exert a rather strong preference for PO_4-ions (in comparison to other anions), indicating the existence of chemical bonds between the solid phase and phosphate. This causes unavailability of phosphates to plant. Addition of fertilizer phosphate to these soils gives usually low response. This phenomenon is called phosphate fixation.

THINK FOR A WHILE

- What is the difference between cation and anion exchange?
- What is the importance of humus and clay particles in cation exchange?
- Why pH is so important in CEC data?
- Why fertile soils contain large number of negative charge?
- How does valence and degree of hydration affect adsorption energy of ions?
- Why phosphate ion is not considered "exchangeable"?

REFERENCES

Bache, B.W. (2006). Ion Exchange. Taylor & Francis.
Bohn, H.L., McNeal, B.L. and O'Connor, G.A. (2001). Soil Chemistry. John Wiley & Sons, Inc.
Bolt, G.H. and Bruggenwert, M.G.M. (1978). Soil Chemistry, A Basic Element. Elsevier Publishers.
Essingtion, M.E. (2005). Soil and Water Chemistry, An Integrative Approach. CRC Press.
Foth, H.D. (1990). Fundamentals of Soil Science. John Wiley & Sons.
McKenzie, N.J., Jacquier, D.J., Isbell, R.F. and Brown, K.L. (2004) Australian Soils and Landscapes An Illustrated Compendium. CSIRO Publishing: Collingwood, Victoria.
Spark, D.L. (2003). Environmental Soil Chemistry. Academic Press.

Chapter 15
pH, Salinity and Sodicity

The pH describes the chemical condition of the soil in terms of acidity and alkalinity and is a clue to treat the soil for necessary requirement. It gives a very significant indication about the soil status. Danish biochemist S.P.L. Sorenson introduced the system of pH for expressing the acidity or alkalinity of solutions.

Water molecules ionize into hydrogen ions and hydroxyl ions. The acidity or alkalinity of the of soil system (including water and other things) is determined by its content of hydrogen and hydroxyl ions. Due to instant chemical reactions, these ions will be absorbed fully/partially in the system. This alters the available positive/negative ions and decide the acidity or basicity of it. Solutions with equal number of hydrogen and hydroxyl ions define the neutral system (Fig. 15.1).

The pH is the negative logarithm of the hydrogen ion activity, i.e. a solution of pH 5 has 10 times the hydrogen ion concentration of a solution of pH 6 and a solution of pH 4 has 100 times the hydrogen ion concentration of a solution of pH 6.

Salinity is the concentration of soluble mineral salts present in soils on a unit volume or weight basis (Page et al., 1982), and sodicity is the concentration of Na^+ ions in soil. The relationship of salinity, sodicity and pH is complex and depends on soil type, concentration of minerals etc. For example, in alkaline soil pH usually increases with increase in salinity due to the presence of sodium bicarbonate and carbonates but pH does not increase necessarily with increase in sodicity. Sodic soils can be neutral or acidic in reaction. Sodic soils of pH 10 is caused by alkalinization with the formation of sodium carbonate and bicarbonate salts. The case is reversed in case of calcareous soil.

DETERMINATION OF SOIL pH

Soil pH is commonly determined by mixing one part of soil with two parts of distilled water or neutral salt solution. Mixing soil and water over a period of 30 min. to allow the mixture to approach an equilibrium condition and then

S. Mukherjee, *Current Topics in Soil Science*,
https://doi.org/10.1007/978-3-030-92669-4_15

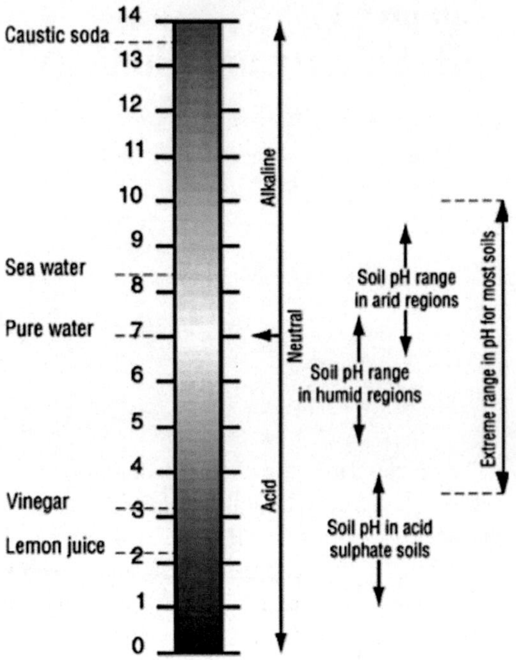

Fig. 15.1 Range of pH values found in soils.

measuring the pH of the soil-water suspension using a pH meter. Normally soil pH ranges from 4 to 10.

SOURCES OF ALKALINITY

Parent rocks from its mineral composition decides the pH of young soils, namely calcareous, mafic soils etc. For instance calcareous soils originate from calcium carbonate rocks and decides its salinity/pH and interactions with acids.

- Hydrolysis decides the alkalinity of soil. For calcium carbonate pH is in the range of 7.6–8.3 and for sodium carbonate pH is above 8.3. This indicates high solubility of Na than Ca.
- Dispersion of clay particles occurs when forces that bind clay particles together are disrupted when too many large sodium ions come between them. Separation of clay particles results in swelling and dispersion which in-turn results in reduced soil permeability due to clogging of soil pores by the clay particles. Although calcium and magnesium contribute to salinity, they generally keep soil flocculated because they compete for the same space as sodium to bind to clay particles. Presence of organic matter can also affect the clogging process. Three main problems caused by sodium-induced dispersion are reduced infiltration, reduced hydraulic conductivity, and surface crusting.

- Weathering of some primary minerals can contribute to alkalinity of soil. For example, weathering of wollastonite ($CaSiO_3$), a pyroxene group mineral produces calcium carbonate that increases the alkalinity in soil

$$CaSiO_3 + CO_2 + 2H_2O \text{à} CaCO_3 \downarrow + H_4SiO_4$$

- When CEC is entirely saturated with calcium, magnesium, potassium and sodium, soil tends to be alkaline. Leaching of the basic cations (Ca, Mg, K and Na) results in their replacement by Al^+ and H^+, and soils becomes acidic. The cations Al^{3+} and H^+ are called the acidic cations.

SOIL ACIDITY

Sources of soil acidity, i.e. addition of H^+, can be done in three major ways:

- Respiration by roots and other soil organisms produces carbon dioxide that reacts with water to form carbonic acid (H_2CO_3) which is a weak acid. This contributes H^+ to the soil solution.
- Acidity is produced when organic matter is mineralized. Nitrogen and sulphur from mineralized organic acids are oxidized to nitric and sulphuric acid, respectively.
- Precipitation reacts with carbon dioxide of the atmosphere and produces carbonic acid which gives natural precipitation—a pH of about 5.6 (acid rain). The results of these reactions add acidity continuously to soils (Foth, 1990).

As precipitation increases, plant growth increases causing more respiration and organic matter mineralization which increases the acidity of soil. Thus, in humid environment soil is more acidic than in desert environment.

What Soil pH Values Mean?

As mentioned before soil pH indicates rather precisely what is going on in the soil in terms of major chemical reactions. These are discussed below (Table 15.1):

- Soil pH values around 2–3 indicates presence of free mineral acid. Most common acid is H_2SO_4 which results from oxidation of iron sulphides. The oxidation takes place during chemical weathering by exposure of sulphur bearing rocks to atmosphere. The long-term effect of these acids in soil causes dissolution of clay minerals which inhibits plant growth. Neutralizing these soils can be very expensive.

Table 15.1 Diagnostic pH values and the dominant reaction

pH	*Dominant reaction*
2–3	Free mineral acid, usually H_2SO_4
4–5	Presence of trivalent, exchangeable Al
≥5.5	Presence of hydroxy-Al
≥6.0–6.7	An ideal pH for many crops
7.6–8.3	Presence of $CaCO_3$
9–10	Presence of Na_2CO_3

- At pH values of 4–5, trivalent exchangeable aluminum is always present in soils unless they are quite high in organic matter. The pH of these soils are dominated by hydrolysis Al^{+3} ions:

$$Al^{+3} + 6H_2O \rightarrow Al(H_2O)_6^{+3}$$

$$Al(H_2O)_6^{+3} + 2H_2O \rightarrow \left[Al(H_2O)_5(OH)\right]^{2+} + 2\left[H_3O\right]^{+}$$

- Root growth is strongly inhibited by Al^{+3} but the case is not as serious as presence of free mineral acids because plants are not killed instantly and neutralizing the soil is not that costly.
- pH 5.5 and above is dominated by complex, polymerized hydroxy-Al instead of any significant trivalent, exchangeable Al^{+3}. The general formula of the polymerized hydroxyl-Al is $[Al(OH)_x(H_2O)_{6-x}{}^{(3-x)+}]_n$ where n is the average number of Al ions per polymer and $x < 3$. These polymers are retained by soil colloids and are nonexchangeable (with few exceptions). Retention of the positive charged polymers lowers the net negative charge of soil colloids decreasing the cation exchange capacity (CEC) in-turn decreasing pH of the soil. Hydroxy-Al is not directly toxic to plants as trivalent-Al but it does have a negative effect of short-term phosphorus availability because it represents a sink for soluble phosphorus.
- A soil with pH 6–7 is free of trivalent exchangeable aluminum, has variable levels of non-exchangeable hydroxy-Al, and has Ca as the dominant exchangeable cation. Manganese, zinc, boron, and iron are usually available in amounts adequate for good crop growth. The 6–7 pH range increases molybdenum availability, which, in-turn, favours crops that depend on nitrogen fixation, such as soybean and alfalfa. This pH also favours mineralization of organic nitrogen with the eventual formation of nitrate. Hence, the pH range of 6–7 generally indicates optimum soil conditions for crop growth, with the fewest negative consequences (Thomas, 2006).
- In pH range of 7.6–8.3 excess $CaCO_3$ is present in the soil. Details of these conditions of alkaline soil have already been discussed in sources of alkalinity.
- As pH values rise above pH 8.3, presence of sodium carbonate is indicated in the soil which has been discussed in sources of alkalinity.

SALINE AND SODIC SOILS

Earth's land surface contains 13.2×10^9 ha of area, out of which 7×10^9 ha is arable and only 1.5×10^9 ha is cultivated. Of the cultivated land 23% is saline and 37% is sodic, containing excessive levels of Na^+. Salt affected soils are found in those areas where insufficient annual precipitation could not meet the evapotranspiration needs of plants. As a result, salts are not leached from the soil, instead they accumulate to inhibit the growth of plants. These salts may include chlorides, sulphates, carbonates and bicarbonates of sodium, potassium, magnesium, and calcium.

Salt-affected soils can be classified as saline, sodic, and saline-sodic. Briefly, saline soils are plagued by high levels of soluble salts, sodic soils have high levels of exchangeable sodium, and saline-sodic soils have high contents of both soluble salts and exchangeable sodium (Sparks, 2003). Table 15.2 shows the global distribution of salt-affected soils.

Salt-affected soils occur mostly in arid and semiarid climates but they can also develop in humid and semihumid regions under appropriate conditions where the climate and mobility of salts cause saline waters and soils for short periods. Some salt-affected soils may occur along sea coasts or river delta regions where seawater has inundated the soil (Richards, 1954).

As plant growth is severely inhibited by the increase of concentration of soluble salts in the soil, the maintenance of agricultural production in these regions necessitates a set of measures. Although these measures will be discussed later in the chapter, it is pointed out that a need for an appropriate 'sink' for the leached salts is required as otherwise the salt removed from one location will move with the groundwater to somewhere else, which simply implies shifting the problem from one spot to another.

Table 15.2 Global distribution of salt-affected soils

Continent	*Area (millions of ha)*		
	Saline	*Sodic (alkali)*	*Total*
North America	6.2	9.6	15.8
Central America	2.0	–	2.0
South America	69.4	59.6	129.0
Africa	53.5	27.0	80.5
South Asia	83.3	1.8	85.1
Southeast Asia	91.6	120.1	211.7
North and Central Asia	20.0	–	20.0
Australia	17.4	340.0	357.4
Europe	7.8	22.9	30.7
Total	351.5	581.0	932.2

Source: I. Szabolcs (1979). Review of research on salt affected soils, UNESCO, Paris

SOURCES OF SOIL SALINITY

The major sources of soluble salts in soils are weathering of primary minerals and native rocks, residual fossil salts (remnants from former marine or lacustrine environments) and atmospheric precipitation. Human activities can also contribute to salinity of soils like irrigation and saline industrial waste. Seawater encroachment also harm soils.

- Exposed rocks and minerals of earth crust releases salts during physical and chemical weathering by the processes of hydrolysis, hydration, oxidation, and carbonation. These salts are either carried down through the soil profile by percolating groundwater during high precipitation or accumulated locally due to scarcity of rainfall.
- The primary source of soluble salts is fossil salts derived from prior salt deposits or from entrapped solutions found in earlier marine or lacustrine sediments. For example, The Lake Mead reservoir in Nevada lies upon deposits of gypsiferous sediments which substantially increased the salinity of Colorado river during passage through the reservoir.
- Significant salt can be deposited from atmosphere. It ranges from 100 to 200 kg $year^{-1}$ ha^{-1} along seacoasts but rapidly changes from 10 to 20 kg $year^{-1}$ ha^{-1} in interior areas of low rainfall. The composition of the salt varies with distance from the source. At the coast it is primarily NaCl whereas the salts become higher in Ca^{2+} and Mg^{2+} further inland.
- Many salt affected soils are a result of human activities. Salts are mainly transported from areas of over-irrigation to accumulate in poorly drainage areas. As the drainage water evaporates, large concentration of salt is left behind. Other human activities like additions of inorganic and organic fertilizers, sludges and sewage effluents, brines from oil and gas fields and mining can also add to salinity of soils.

SALINITY PARAMETERS

Total Dissolved Solids (TDS)

Total dissolved solids (TDS) can be measured by evaporating a known volume of water from the solid material to dryness and weighing the residue. However, this measurement is variable since in a particular sample various salts exist in varying hydration states, depending on the amount of drying. Thus, if different conditions are employed, different values for TDS will result. TDS is a useful parameter for measuring the osmotic potential $-\boldsymbol{\tau}_o$, an index of the salt tolerance of crops are in the range of 5–1000 mg $litre^{-1}$ TDS (Bresler et al., 1982).

Electrical Conductivity (EC)

The most important parameter required for determining soil salinity and sodicity is electrical conductivity (EC), and the concentrations of calcium, magnesium, and sodium. The ability of a solution to conduct electricity is directly related to the concentration of dissolved salts.

The EC is determined by placing two flat, rectangular electrodes that are separated by a constant distance into a solution. These electrodes are contained in a conductivity cell. A constant potential is applied to the electrode and a current flown through the solution between the electrodes (Essington, 2005).

A soil is saline if the EC of the saturation extract (EC_e) is higher than 4 dSm^{-1} (decisiemens per metre). However, plant growth may be severely affected or inhibited at much lower levels of EC whose values change with the ionic composition of the soil solution and with the plant species and variety grown.

For salt-sensitive plants, $EC_e > 2$ dSm^{-1} would denote a saline soil; whereas, for salt-tolerant plants, a saturation extract $EC_e > 8$ dSm^{-1} would denote a saline soil. EC_e levels have also been defined to indicate the types of crops that can tolerate various levels of soil salinity. Saline soils are defined as $EC_e > 1.5$ dSm^{-1} for sensitive crops; $EC_e > 3.0$ dSm^{-1} for moderately sensitive crops; $EC_e > 6$ dSm^{-1} for moderately tolerant crops; and $EC_e > 10$ dSm^{-1} for tolerant crops (see Maas, 1990).

SODICITY PARAMETERS

Sodic soils are those with a high proportion of sodium adsorbed to the soil particles compared to other cations such as Ca, Mg, and K.

Sodium adsorption ratio (SAR), the *exchangeable sodium ratio (ESR),* and the *exchangeable sodium percentage (ESP)* are the most important parameters for measuring sodicity of soils. SAR is commonly measured using the equation

$$\text{SAR} = \left[\text{Na}+\right]/\left[\text{Ca}^{2+}+\text{Mg}^{2+}\right]^{1/2}$$

and has many limiting factors as

- it treats Ca^{2+} and Mg^{2+} as same species but in many waters, concentration of Ca^{2+} is much higher than Mg^{2+}.
- SAR uses total concentrations, not activities, and thus, it does not consider decreases in free ion concentrations and activities due to ion pair or complex formation, which can be significant with Ca^{2+} and Mg^{2+}.

SAR can be simplified, since Na^+, Ca^{2+}, and Mg^{2+} are the most common exchangeable ions in arid soils (Jurinak and Suarez, 1990) to

$$[\text{Na}-\text{Soil}]/\text{CEC}-[\text{Na}-\text{Soil}]=\text{k'G}\,\text{SAR}=\text{ESR}$$

where the concentration of the ions on the exchanger phase and CEC are expressed in mole kg^{-1}, *k′G* is the modified Gapon selectivity coefficient, and ESR the exchangeable Na^+ ratio (Richards, 1954). The *k′G* (mmol $litre^{-1})^{-1/2}$, is expressed by

$$\text{k'G}=[\text{Na}-\text{Soil}]\left[\text{Ca}^{2+}+\text{Mg}^{2+}\right]^{1/2}/[\text{Ca}-\text{Soil}+\text{Mg}-\text{Soil}]\left[\text{Na}^{+}\right]$$

Traditionally, a soil is by definition sodic if the ESP is greater than 15 (United States Salinity Laboratory Staff, 1954):

$$\text{ESP}=100\text{X}(\text{ESR}/1+\text{ESR})$$

Sodic soils adversely affect soil structure and movement of air and water into the soil and to plant roots thereby indirectly affecting plant growth and survival whereas soil salinity reduces plant growth, directly affecting physiological functions through osmotic and toxicity effects on plants. Excess sodium on the exchange complex causes clay dispersion and swelling which subsequently clog soil pores and reduce soil hydraulic conductivity (Essington, 2005). Classification of saline and sodic soil depending on these parameters are in Table 15.3).

RECLAMATION

Saline soil contains soluble salts like Cl^-, SO_2^{-4} and sometimes NO^{-3}. But because exchangeable Na^+ is not a problem, saline soils are usually flocculated and water permeability is good. Saline soil requires the movement of salts out of the profile to a depth that is well below the root zone. Reclamation is accomplished by leaching. Leaching the soil with good-quality (low electrolyte concentration) water dissolves the salts and removes them from the root zone. Most irrigation waters are suitable for reclaiming saline soils but the amount of water required to achieve reclamation will be a function of initial soil salinity, the soil type (texture), depth of reclamation, and the technique for applying water. Reclamation can be hampered by several factors such as restricted drainage caused by a high groundwater table, low soil

Table 15.3 General classification of salt affected soils

Soil class	*EC* (dSm^{-1})	*SAR*	*ESP (%)*
Nonsaline nonsodic	<4	<13	<15
Saline	>4	<13	<15
Saline-sodic	>4	>13	>15
Sodic	<4	>13	>15

hydraulic conductivity due to restrictive soil layers, lack of good-quality water, and the high cost of good quality water (Bresler et al., 1982).

For sodic soils Na^+ is the major problem. The high amount of Na^+ in these soils, along with low EC_e, results in dispersion. Thus, reclamation requires adequate drainage, leaching, and a source of Ca^{2+}. The Ca^{2+} can be acquired by applying gypsum ($CaSO_4 \cdot 2H_2O$) or $CaCl_2$ to remove the exchangeable Na^+. The Ca^{2+} exchanges with Na^+, which is then leached out as a soluble salt Na_2SO_4 or NaCl. Gypsum or $CaCl_2$ can also increase permeability by increasing electrolyte concentration. If calcite is present in the soil, H_2SO_4 can be used to dissolve calcite.

Saltwater-dilution method is also effective in reclamation in which soil is rapidly leached with water that has a high electrolyte concentration (low SAR water) having large quantities of Ca^{2+} and Mg^{2+}. Highly saline, low SAR water increases the permeability of the soil. After leaching, and removal of Na^+ from the exchanger phase of the soil, the soil is leached with water of lower electrolyte concentration to remove excess salt.

Green manuring as a reclamation practice consists of incorporating plant residues into the soil and leaching with water. The high CO_2 production is combined with restricted gas transport creating very high CO_2 concentrations in the soil, dissolving large amounts of calcium carbonate. It is estimated that this process can dissolve in the order of 400–800 kg ha^{-1} during a year of reclamation.

CONCLUSIONS

Soil pH is one of the most useful, economic and rapid measurements that can be made on soils. Different pH values tell different dominant reactions occurring in soil. To assure plant growth without serious inhibition from unfavourable soil pH, right plants can be selected for existing soil pH, or pH of the soil can be altered to suit the needs of the plants. Most soil pH changes are directed towards reduced soil acidity and increased pH by liming. Salt-affected soils decrease crop productivity, both locally and globally. Thus, problem is likely to increase if no measures are taken to limit soil salinization and sodication. Management and reclamation of salt-affected soils require good agronomic practices. Increase in improved crop salt tolerance establishes a proper water balance and reduce the need and cost of reclamation measures. Information on crop response to salinity and sodicity needs to be updated in relation to changes of plant materials, agronomic practices and climatic conditions.

THINK FOR A WHILE

- How does Na ion affect the clogging process of clay minerals?
- What are the different processes that indicates acidity of a soil?
- Why are saline soils found in dry areas?

- What are the primary source of salt in saline soil? Give an example.
- What is the most common method for measuring sodicity of soil? What are its limiting factors?
- How reclamation is achieved in saline soils?

REFERENCES

Bresler, E, McNeal, B.L. and Carter, D.L. (1982). Saline and sodic soils: Principle – dynamics-modelling, vol. 10, Advanced Series in Agricultural Sciences, Springer Verlag, Berlin.

Essingtion, M.E. (2005). Soil and Water Chemistry, An Integrative Approach. CRC Press.

Foth, H.D. (1990). Fundamentals of Soil Science. John Wiley & Sons.

Jurinak, J.J. and Suarez, D.L. (1990). The chemistry of salt affected soils and waters. *In* Tanji, K. K. (ed.) Agricultural Salinity Assessment and Management. American Society of Civil Engineers, New York. pp. 42–63.

Maas, E.V. (1990). 'Crop salt tolerance', in K.K. Tanji (ed.). Agriculture salinity assessment and management, American Society of Civil Engineers, New York, pp. 262–304.

Page, A.L., Miller, R.H. and Keeney, D.R. (1982). Methods of Soil Analysis. Part 2. Chemical and Microbiological Properties. American Society of Agronomy. In Soil Science Society of America, Vol. 1159.

Richards, L.A. (1954). Diagnosis and Improvement of Saline and Alkali Soils. US Salinity Laboratory Staff, US Department of Agriculture, Washington DC.

Spark, D.L. (2003). Environmental Soil Chemistry. Academic Press.

Szabolcs, I. (1979). Review of research on salt-affected soils. Paris, France: UNESCO.

Thomas, G.W. (2006). pH. Taylor & Francis.

Chapter 16
Chemistry of Soil Nutrient

Nutrients are elements or compounds that are essential for the growth and survival of life. Plants require large amounts of nutrients and can be divided in three categories (Fig. 16.1).

Plant growth is hampered with the nutrition deficiency at the same time, too much nutrients makes the system toxic.

In the plant development, the deficiency or toxicity is reflected characteristically by the visual symptoms on leaves, stems, fruit, and roots etc. A careful observation of these visual symptoms specifying certain disorders can control the problem.

Nutrients differ in their mobility and extent of re-translocation. Some nutrients which are readily mobile such as nitrogen, potassium, phosphorus, magnesium are easily re-translocated from older to younger tissues when nutrient supply becomes limiting. Deficiency symptoms of these elements develop first in the older tissues and advance towards the younger tissues as the deficiency becomes more severe. Nutrients that have low mobility (e.g., calcium, boron, iron) develop symptoms first in immature tissues. Mobility can also vary depending on the degree of deficiency, and either the nitrogen or phosphorus status of the plant. For example, copper, zinc, and molybdenum are relatively immobile when nitrogen supply is adequate, and there is little or no movement of these elements from old leaves of plants deficient in them. Here, symptoms occur mainly in young tissues. However, should nitrogen become limited, copper, zinc, and molybdenum can be remobilized from older to younger leaves along with the re-translocation of nitrogen, and the symptoms of nitrogen deficiency are likely to dominate (Grundon, 2006).

S. Mukherjee, *Current Topics in Soil Science*,
https://doi.org/10.1007/978-3-030-92669-4_16

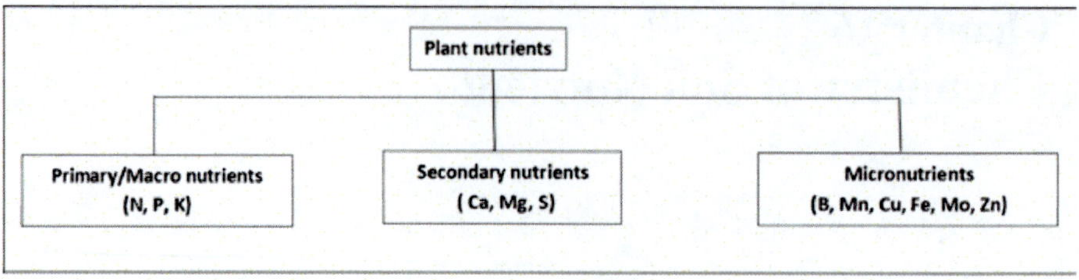

Fig. 16.1 Types of plant nutrients.

Table 16.1 Summary of functions of various primary macronutrients

Plant nutrients	*Functions*
1. Nitrogen (N) (Mostly absorbed from atmosphere)	• Necessary for formation of amino acids, the building blocks of protein • Essential for plant cell division, important for plant growth • Helps in production and use of carbohydrates • Aids in energy reaction in the plants
2. Phosphorus (P) Oxide: Apatite $Ca_5(PO_4)_3(F,Cl,OH)$	• Promotes early root formation and growth • Essential for seed formation • Increases water-use efficiency • Involved in photosynthesis, respiration, energy storage and transfer, cell division • Present within plants cells as structural component of the nucleic acids such as DNA and RNA as well as a constituent of fatty phospholipids, that are important in membrane development and function • It is used in protein synthesis and is found in legume nodules
3. Potassium (K) Chloride: Sylvite (KCl) Sulphate: Langbeinite $K_2Mg_2(SO_4)_3$	• Increases photosynthesis • Important for fruit formation • Increases photosynthesis, water-use efficiency • Carbohydrate metabolism and the breakdown and translocation of starches • Activates enzymes and control their reaction rates • Increase disease resistance

MACRONUTRIENTS

Plants require large amounts of nutrients such as nitrogen (N), phosphorus (P), carbon (C), hydrogen (H), oxygen (O), potassium (K), calcium (Ca), magnesium (Mg) and Sulphur (S). These are called macronutrients that make up the greatest proportion of the plant (Tables 16.1 and 16.2).

Table 16.2 Summary of functions of various secondary macronutrients

Plant nutrients	*Functions*
1. Calcium (Ca) Carbonate: Limestone ($CaCO_3$) Sulphate: Anhydrite ($CaSO_4$) Gypsum ($CaSO_4 \cdot 2H_2O$)	• Calcium attaches to the walls of plant tissues, stabilizing the cell wall and favouring cell wall formation. It is also involved in cell growth and development • It improves plant health by activating the formation of roots and their growth • Calcium contributes to mineral retention and its transportation in the soil. It is also involved in the activation of certain plant enzymes
	• Calcium helps in combating salinity in soils. It has been shown to ameliorate the negative effects that salinity has such as reduced water usage of plants • It neutralizes toxic substances in plants and contributes to seed formation
2. Magnesium (Mg) Carbonate: Dolomite $CaMg(CO_3)_2$ Magnesite ($MgCO_3$) Hydroxide: Brucite $Mg(OH)_2$	• Magnesium constitutes the core of the chlorophyll molecule and is therefore essential for photosynthesis. This makes it an very important element for plant development • Magnesium promotes the absorption and transportation of energy supplying phosphorus compounds • It contributes to the storage of carbohydrates within the plant • Magnesium acts as an enzyme activator, and in fact activates more enzymes than any other nutrient
3. Sulphur (S) Sulphide: Pyrite (FeS_2) Galena (PbS) Sphalerite (ZnS) Sulphate: Gypsum ($CaSO_4 \cdot 2H_2O$)	• Helps to develop enzymes and vitamins • Aids in seed production • Promotes nodule formation on legumes • Necessary for chlorophyll formation • Important constituent of plant proteins

NITROGEN

Nitrogen gas makes up about 78% of earth's atmosphere by volume as inert N_2 and is one of the primary nutrients critical for the survival of all living organisms. Although nitrogen is very abundant in the atmosphere, it is largely inaccessible in this form to most organisms making nitrogen a scarce resource and often limiting primary productivity in many ecosystems. Only when nitrogen is converted from dinitrogen gas into other forms does it become available to primary producers, such as plants.

Nitrogen Cycle

The nitrogen cycle is the biogeochemical cycle by which nitrogen is converted into multiple chemical forms as it circulates among atmosphere, terrestrial, and marine ecosystems. The conversion of nitrogen can be carried out through both biological and physical processes. Soil nitrogen cycle is a sequence of biochemical changes wherein nitrogen is used by living organisms, transformed upon death and decomposition of the organisms, and converted ultimately to its original state of oxidation (Fig. 16.2).

The N_2 molecule has extremely strong triple bonds that resists them to react with other elements. To be used by plants conversion of N_2 is important. The conversion processes are:

Nitrogen Fixation Biological nitrogen fixation is where microbes form symbiotic relationships with plants. The microbes provide N to the plants and the plants provide carbohydrates to the microbes from photosynthesis. The main nitrogen fixing bacteria is *Rhizobium* but other bacteria genuses such as *Frankia* and some species of free-living or lichen-forming cyanobacteria can also fix N. These bacteria have the nitrogenase enzyme that combines gaseous nitrogen with hydrogen to produce ammonia, which is converted by the bacteria into other organic compounds.

Rhizobium lives in symbiotic relationship in root nodules of legumes such as clover, lucerne, peas and beans that converts atmospheric N_2 gas to NH_3, which is further converted to amino acids and proteins. In return, the *Rhizobium* species receive from the legume the energy they need to grow and to fix N_2. There are also non-symbiotic bacterias such as *Azotobacter* that make a small but significant contribution to the soil N status. They depend on decomposable organic matter for their energy supply and thus, have limited activity in soils.

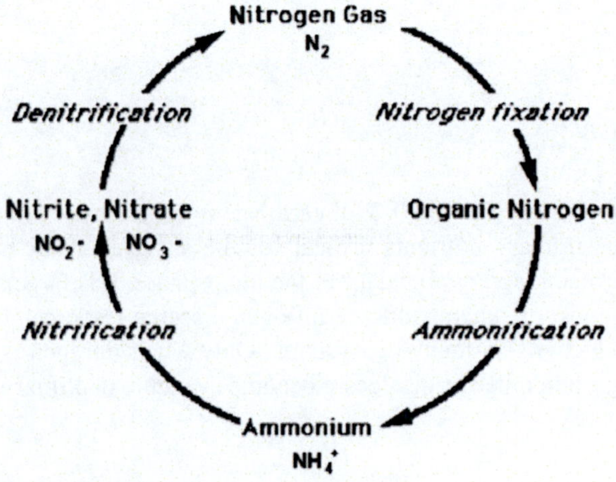

Fig. 16.2 General nitrogen cycle.

- Many non-legume plant species like Red alder have root nodules and can also fix nitrogen symbiotically.
- Some nitrogen can be fixed abiotically by lightning or certain industrial processes such as combustion of fossil fuels. The amount of globally fixed N is almost double the amount of applied fertilizer N.

Ammonification/Mineralization About 1–2% of the total nitrogen in soil is decomposed and mineralized by different kinds of heterotrophic organisms and converted into ammonia. This process is called ammonification.

- NH_3 is released from animal waste or manure, and some NH_3 produced at or near the soil surface escapes into the atmosphere especially when the soil becomes alkaline. Since ammonia is a polar gas, it reacts with water and combines with a H^+ to form ammonium (NH_4^+). Ammonium is adsorbed onto the cation exchange sites and is retained in soils as an exchangeable cation. The ammonium ion is about the same size as the potassium ion, and some NH_4^+ becomes entrapped in the interlayer space of micaceous minerals, somewhat like entrapment of K^+. This entrapment of ammonium nitrogen is called *ammonium fixation*. Fixed NH_4^+ has low plant availability (Foth, 1990).
- Unlike symbiotic plants, nonsymbiotic plants depends totally on soil for their nitrogen needs. Amount of nitrogen a soil can supply to these plants depends on the amount of nitrogen mineralized, which is dependent on the kind and amount of soil organic matter and the rate of nitrogen mineralization.

Nitrification Nitrification is a two step process by which ammonia is converted to nitrite (NO_2) and then to nitrate (NO_3^-) by biological oxidation. They are carried out by specific autotrophic bacterias like *Nitrosomas* and *Nitrobacter*. The reactions are as follows:

$$NH_4^+ + 1.5O_2\,(\text{Nitrosomas}) \rightarrow NO_2^- + 2H^+ + H_2O + \text{Energy}$$

$$NO_2^- + 0.5O_2\,(\text{Nitrobacter}) \rightarrow NO_3^- + \text{Energy}$$

In the first process of nitrification H^+ is released which contributes to the acidity of the soil. Nitrification is generally affected by oxygen supply and pH. It does not take place if oxygen is absent (water saturated soil). That is why paddy rice is more productive when an ammonium fertilizer is applied rather than a nitrate fertilizer. Nitrifying bacteria thrives in well aerated soils with pH 6 or more. Hence, a neutral pH is favourable for nitrification.

Since nitrate is highly water soluble it remains in soil solution and is readily absorbed by the plant roots compared to ammonia which gets adsorbed onto the cation exchange sites and resists leaching. Nitrate leaching can cause serious groundwater pollution and that is why ammonium is used as a fertilizer during high rainfall periods.

Immobilization Ammonia and nitrate are forms of nitrogen used by plants and microorganisms. When there is a shortage of nitrogen in soil, plants and microorganisms have to compete for it. The process in which nitrogen is taken up by soil organisms and therefore become unavailable to crops is called immobilization. It is a process by which inorganic nitrogen is converted to organic form by plants and soil organisms. Therefore it is opposite to mineralization. In fact they happen continuously and concurrently in nature.

Denitrification The process of molecular nitrogen being released back to the atmosphere by reduction from nitrate or nitrite is called denitrification. Since nitrogen is fixed in the ecosystem, there is a balance between nitrogen fixation and denitrification.

- Denitrification is carried out by some anaerobic organisms. It mainly favours anoxic conditions and poorly drained soils (anaerobic conditions may exist in well-drained soils after rains, when the interiors of soil aggregates are wet). Denitrification helps in preventing groundwater pollution by removing excess of nitrate from soil. Thus, it is mainly used to remove nitrogen from sewage and municipal wastewater ((Fig. 16.3).

PHOSPHORUS

Among the primary macronutrients like nitrogen and potassium, phosphorus occurs in the least amount in soils and is the most growth-limiting factor. Nevertheless it is one of the most vital element that support life on earth. The major problem in phosphorus uptake by plants is the very low solubility of most phosphorus compounds, resulting in a low concentration of phosphate ions in the soil solution.

Phosphorus Cycle

The biogeochemical cycle that describes the movement of phosphorus through the lithosphere, hydrosphere, and biosphere is called the phosphorus cycle (Fig. 16.4). Unlike many other biogeochemical cycles, the atmosphere does not play a significant role in the movement of phosphorus because phosphorus in soil is mainly found in solid compound form.

The earths crust contains about 0.1% of phosphorus (P) which is mostly found in the mineral apatite. Apatite weathers slowly and phosphorus is mainly released in soil as $H_2PO_4^-$ (when soil pH is below 7.2) and as HPO_4^{2-} (when soil pH is above 7.2).

Following steps are of phosphorus cycle:

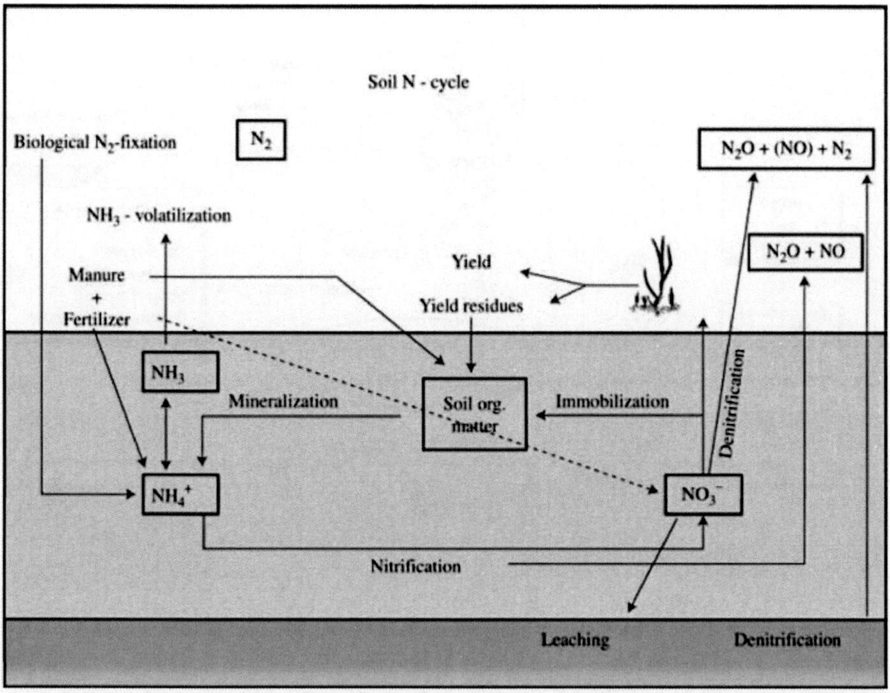

Fig. 16.3 The soil N cycle. White compartment represents atmosphere; light gray represents the biosphere and the dark gray subsoil.

- Weathering caused the minerals to release phosphorus as $H_2PO_4^-$ and HPO_4^{2-} into the soil.
- Like nitrogen *immobilization* of inorganic P takes place by the roots and micro-organisms when they absorb and convert it into organic P. Opposite process i.e. *mineralization* also takes place by the microorganisms to convert the organic P back to inorganic form. This is a subcycle that takes place with the overall P cycle.
- *Phosphorus Fixation:* The major difference between nitrogen and phosphorus cycle is that in N cycle, nitrogen is mainly fixed as nitrate which is stable in soil solution and is taken by plants but in P cycle, phosphate ion quickly reacts with other ions in the soil solution, resulting in precipitation and adsorption to mineral colloids that convert the phosphorus to an unavailable or fixed form. This is called phosphorus fixation. This causes phosphorus unavailability to plants and is mainly dependent on soil pH.

In calcareous soil, soil solution has excess Ca^{2+}. Any phosphate ion released will precipitate as calcium phosphate which slowly will convert to apatite. Thus, phosphorus in such condition remains as apatite and plant availability of P is low.

In humid regions, where there is excess rainfall, calcium is leached and the soil becomes acidic. This causes an increase of Al^{+3} and Fe^{+2} ions. Phosphate ions release

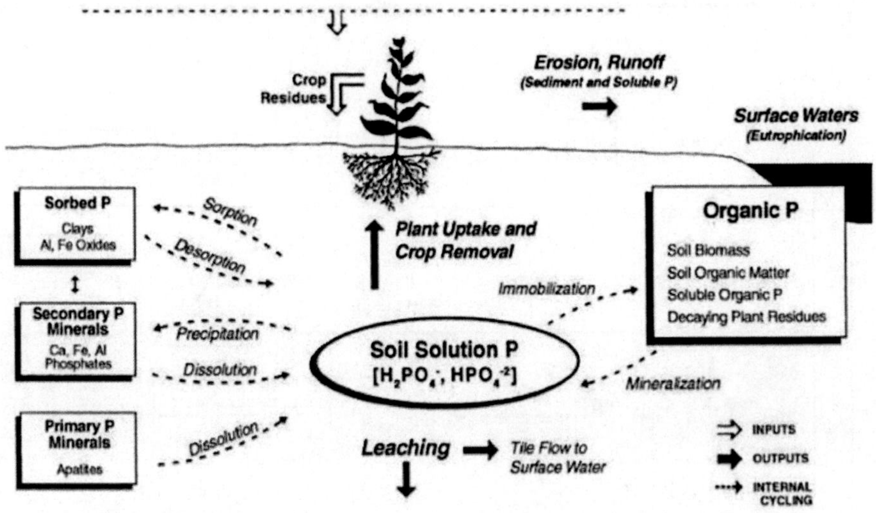

Fig. 16.4 Soil phosphorus cycle.

will precipitate into insoluble aluminum and iron phosphates which is again unavailable to plants.

- Dissolution or desorption of phosphorus takes place but very slowly. Optimum pH of 6.5 will cause minimum phosphorus fixation but the amount of phosphorus in soil solution will be very less.

Hence, to avoid this problem, phosphorus fertilizers are used in crops. Phosphorus fertilizer fixation takes place and only 10 to 20% of phosphorus in fertilizer is used by plants during the year of fertilizer application. Therefore, repeated use of phosphorus fertilizers is required for an increase in soil phosphorus content.

POTASSIUM

Potassium (K) is one of the most important macronutrient for plant growth and development. After nitrogen, K has the largest concentration in plant ranging from 5 to 50 g kg^{-1}. Unlike N, P and S, potassium is present in the soil solution only as a positively charged cation (K^+). It is involved in a large number of physiological processes such as osmoregulation and cation-anion balance, protein synthesis, carbohydrate metabolism and activation of enzymes. Potassium is essential for photosynthesis, N-fixation in legumes, starch formation, and translocation of sugars. It also regulates the opening and closing of plant stomata by reducing water loss from the leaves and increasing its drought tolerance. Potassium has an important role in helping plants adapt to winter hardiness, fungal diseases and insect pests.

Potassium Cycle

The earth crust contains an average of 2.6% of potassium which in soil comes from minerals containing it. During weathering these minerals release K^+ ions which are then adsorbed by the cation exchange sites. The exchangeable potassium tends to maintain an equilibrium concentration with the soil solution from which roots absorb K^+.

Minerals like feldspar (microcline, orthoclase) and micas (muscovite and biotite) mainly contains K. Micas tends to weather faster than feldspar and thus micas are found among smaller particles (silt and clay) than feldspars (sand).

Potassium cycle occurs in following steps (Fig. 16.5):

- During weathering, when the minerals releases potassium into the soil, it mainly goes into the soil solution or adsorbed into cation exchange sites.
- Soon an equilibrium tends to establish between K^+ of soil solution and exchange sites.
- When weathering continues, there is no removal of K^+ from soil solution by plant uptake or leaching (during winter), K^+ in the exchangeable sites increases. This causes fixation of potassium where K^+ moves back into voids of the minerals where interlayer potassium weathered out at an earlier time.
- The opposite happens when leaching and plants uptake causes removal of K^+ from soil solution (spring and summer) and thus decreases K^+ in the exchangeable sites. This causes release of fixed potassium to the soil.

Potassium concentration varies from soil to soil. Since K does not complex with organic compounds, organic soils are mostly potassium deficient. Soils with abundant mica and vermiculite tends to have the highest potassium concentration.

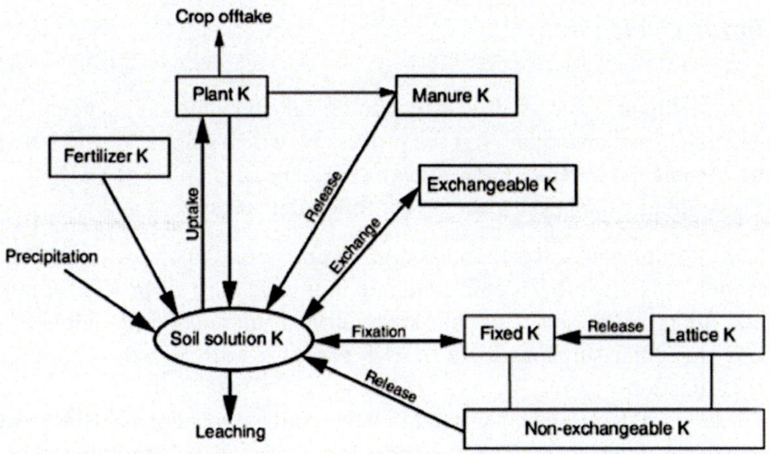

Fig. 16.5 Potassium cycle.

CALCIUM AND MAGNESIUM

Calcium and magnesium are secondary macronutrients unlike N, P, K and are required less by the plants compared to former nutrients. Like K^+, calcium and magnesium are taken up by plants as cation Ca^{+2} and Mg^{+2}, respectively. Ca^{+2} and Mg^{+2} too were released from minerals by weathering and occur as exchangeable cations. Calcium minerals include calcite, dolomite, gypsum, feldspar, apatite, and amphibole whereas magnesium minerals are dolomite, biotite, serpentine, hornblende, and olivine.

Calcium and magnesium cycles are rather simple since it does not involve fixation or release. Like potassium, weathering causes release of Ca^{+2} and Mg^{+2} into soils solution and cation exchange sites until they reach an equilibrium. Compared to potassium, the amounts of exchangeable and solution calcium and magnesium are usually many times larger but their needs for plants is much lesser than their supply. Thus, deficiency of calcium and magnesium in plants are very rare. Magnesium deficiency is usually associated with acidic sandy soils whereas calcium deficiency occurs on intensively weathered soils.

SULPHUR

Sulphur is an essential element for growth and physiological functioning of plants. It is mainly found in sedimentary rocks like gypsum and coal and in very low concentration in igneous and metamorphic rocks. Mineral weathering releases sulphur as sulphate (SO_4^{2-}) which is absorbed by roots and microorganisms. Some plants also take atmospheric sulphur as sulphur dioxide (SO_2) through stomata.

Sulphur Cycle

The soil sulphur cycle is driven by biological and physico-chemical processes, which affect flora and fauna. It is the process by which sulphur moves between soil, living organisms, water and atmosphere (Fig. 16.6).

The sulphur cycle takes place by the following steps:

- 95% of sulphur in soil is found as organic compound (mainly cysteine and methionine). This organic S is converted into inorganic S by the process of mineralization. Inorganic forms of sulphur are hydrogen sulphide (H_2S), sulphide minerals and elemental sulphur. Some of H_2S can also be released to atmosphere via volatilization.
- These inorganic forms of sulphur is then oxidized to sulphate (SO_4^{-2}) by microorganisms such as some photosynthetic bacterias and chemolithotrophs and are taken by plants. Opposite process happens (immobilization) when there is a

shortage in S in soil thus plants and microorganisms convert the inorganic S back to its organic form.

- Sulphate is then reduced to sulphide by suphate reducing microorganims.

Although sulphur is needed by plants as much as phosphorus, it is rarely deficient in plants. Sulphur deficiencies tend to occur where there is little input of sulphur by precipitation (during low precipitation) and little input of sulphur into the atmosphere by burning of fossil fuels. Thus concentration of sulphur in soil is really high in industrious areas. Addition of sulphur decreases the pH of soil thus increasing the capacity for sulphur adsorption.

MICRONUTRIENTS

Micronutrients are needed in small quantities, but are no less important; however, deficiencies of these are less likely to occur. They are boron (B), manganese (Mn), iron (Fe), copper (Cu), zinc (Zn) and molybdenum (Mo) (Table 16.3).

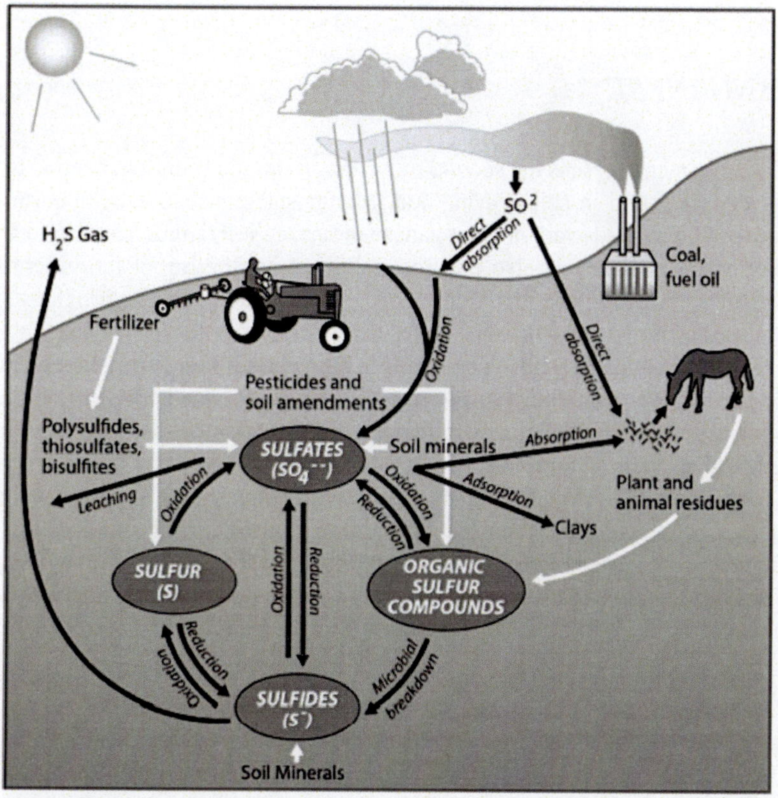

Fig. 16.6 Sulphur cycle.

IRON

Iron is the fourth most abundant element in earth's crust following O, Si and Al and comprises about 5% of earth's crust. Iron in soil is mainly due to weathering of its minerals such as its oxides (haematite, magnetite, goethite), sulphides (pyrite), and silicates (olivine).

Of all micronutrients, plants needs iron the most. Iron content in soil varies from 0.4 to 27.3% whereas plant available iron content in soil ranges from 0.36 to 174 ppm. However, solubility of iron in soil depends mainly on pH. Iron is least soluble when pH of soil is within 7.4 to 8.5 causing iron deficiency in plants. With increase in soil pH oxidized forms of iron exist as insoluble oxides and hydroxides. Thus, plants have developed special coping mechanisms in alkaline soils to deal with the deficiency:

- Plant decreases pH of its roots by excretion of H^+ which dissolves the insoluble iron oxides.
- Few plants (monocots) excretes iron bodies called siderophores which solubilizes iron oxides. Some microorganisms too can excrete siderophores which satisfy needs of the plants.

MANGANESE

After iron, manganese is the second most important micronutrient needed by plants. It is mainly found in soil as Mn^{+2} ion. Manganese concentration in earth crust is about 1000 ppm and available manganese in surface soil ranges from 0.6 to 164 ppm. Like iron, manganese is also present in soil by weathering of its minerals. Main manganese minerals are its oxides (pyrolusite, psilomelane), hydroxides (manganite), carbonates (rhodochrosite) and silicate (rhodonite).

Iron and manganese has very similar characteristics and so its deficiencies are also similar. It is sometimes impossible to understand whether the symptom is due to iron or manganese deficiency unless trials are conducted to determine which element alleviates the symptoms.

Manganese concentration too decreases with increase in pH. At pH 7 the microbial activity reaches maximum where microorganisms in soil oxidizes the soluble manganese into unavailable oxide forms. Soil texture and organic matter too can affect the Mn concentration. Coarse textured soil rich in organic matter can decrease Mn concentration owing to formation of chelated Mn^{+2}. Lastly, lower redox potential (waterlogged soils) favours increase in Mn concentration.

Table 16.3 Summary of functions of various micronutrients

Plant nutrient	*Functions*
1. Iron (Fe) Oxides: Haematite (Fe_2O_3) Magnetite (Fe_3O_4) Goethite FeO(OH) Sulphides: Pyrite (FeS_2) Silicate: Olivine (Mg, Fe)$_2$SiO$_4$	• Iron helps in chlorophyll synthesis and thus its deficiency can cause chlorosis • It serves as a component of many vital enzymes such as cytochromes of the electron transport chain, and it is thus required for a wide range of biological and physiological functions • Iron plays a critical role in metabolic processes such as DNA synthesis, respiration, and photosynthesis • It also activates nitrogen fixation
2. Manganese (Mn) Oxides: Pyrolusite (MnO_2) Psilomelane $(Ba,H_2O)_2Mn_5O_{10}$ Hydroxides: Manganite MnO(OH) Carbonate: Rhodochrosite ($MnCO_3$) Silicate: Rhodonite $(Mn^{2+},Fe^{2+},Mg,Ca)SiO_3$	• Manganese plays an important role in oxidation and reduction processes in plants, such as the electron transport in photosynthesis • Manganese is a part of about 35 multiple enzymes and act as a catalyst on many other enzymes. So it has a metabolic role in the nitrate-reducing enzyme activity and activation of enzymes which play important roles on carbohydrate metabolism • It is also involved in pollen germination, pollen tube growth, root cell elongation and resistance to root pathogens
	• Manganese plays a key role in photosynthesis. It facilitates the photolysis of water molecules and provides energy for photosynthesis • It contributes to abiotic stress tolerance, particularly drought and heat • Increases the availability of P and Ca
3. Zinc (Zn) Sulphide: Sphalerite(Zn,Fe)S Carbonate: Smithsonite ($ZnCO_3$) Silicate: Hemimorphite $Zn_3Si_2O_7(OH)_2 \cdot H_2O$	• Zinc activates enzymes that are responsible for the synthesis of certain proteins and photosynthesis • It causes metabolism of carbohydrate, lipid and nucleic acid. Zn is part of the synthesis of the plant hormone auxin which help with growth regulation and stem elongation • Zn is also key for seed and grain maturation and production and its presence in plant tissue helps the plant to withstand cold temperatures

(continued)

Table 16.3 (continued)

Plant nutrient	*Functions*
4. Copper (Cu) Oxides: Cuprite (Cu_2O) Sulphides: Chalcocite (Cu_2S) Covellite (CuS) Carbonate: Malachite $Cu_2CO_3(OH)_2$ Azurite $Cu_3(CO_3)_2(OH)_2$ Silicate: Chrysocolla $(Cu,Al)_2H_2Si_2O_5(OH)_4{\cdot}n(H_2O)$	• Copper is an essential element for plant growth. It activates some enzymes in plants which are involved in lignin synthesis and it is essential in several enzyme systems. Lignin synthesis is needed for cell wall strength and prevention from wilting • It is required in the process of photosynthesis and Cu-containing proteins are also important in respiration (combusting sugars for energy in plants) • It is used to intensify flavour and colour in vegetables and colour in flowers
5. Boron (B) Oxides: Borax ($Na_2B_4O_7{\cdot}10H_2O$) Kernite ($Na_2B_4O_6(OH)_2{\cdot}3H_2O$) Colemanite ($Ca_2B_6O_{11}{\cdot}5H_2O$)	• Boron is used with calcium in cell wall synthesis and is essential for cell division (creating new plant cells) • Boron plays the most important role in reproductive growth so it helps with pollination, and fruit and seed development • Other functions include translocation of sugars and carbohydrates, regulating nitrogen metabolism, formation of certain proteins
Plant nutrient	*Functions*
	• Boron also helps in transportation of potassium to stomata which helps regulate internal water balance • Boron contributes more than any other micronutrient to the quality of produce
6. Molybdenum (Mo) Sulphide: Molybdenite (MoS_2) Oxide: Ilsemmannite (Mo_3O_8) Wulfenite ($PbMoO_4$) Powellite ($CaMoO_4$)	• Molybdenum is an essential component in enzymes that convert nitrate into nitrite and then into ammonia before using as a synthesizer of amino acids within the plant • It is also needed by symbiotic nitrogen fixing bacteria in legumes to fix atmospheric nitrogen • Plants also use molybdenum to convert inorganic phosphorus into organic forms in the plant

ZINC

Zinc content of earth's crust is 80 ppm and plant available zinc content in soil varies from 0.8 to 20.5 ppm. Zinc is released from mineral as Zn^{+2} and it includes sulphides (sphalerite), carbonate (smithsonote) and silicates (hemimorphite).

Zinc is water soluble, immobile, act as a exchangeable cation and adsorbed to surfaces of organic matter, clay (bentonite, illite and kaolinite) and other minerals (magnesite, calcite, dolomite). It also forms complexes with organic matter and substitutes for Mg^{+2} of clay minerals.

Like all micronutrients, Zn availability too decreases with pH. Zn^{+2} predominates below pH 7.7 whereas above hydrolysis of zinc takes place to form hydroxides which are highly insoluble.

Submerged soils raises the pH of acidic soil but lowers the pH of alkalinite soil. Thus, in acidic submerged soil Zn availability decreases whereas it increases for alkaline soil. This is due to the formation of franklinite ($ZnFe_2O_4$) which is a soluble form of zinc mineral.

Some cations such as Cu^{+2}, Fe^{+2} and Mn^{+2} inhibit the uptake of Zn by plants due to competition for the same site whereas solubility and mobility of Zn increases in presence of SO_4^{-2}.

COPPER

Copper content in earth's crust is about 55–70 ppm and plant available copper concentration in soil ranges from trace to 32 ppm. Although copper content in soil is low, plants need very small amount of Cu and thus its deficiency rarely takes place.

Sulphides are the most common minerals of copper (chalcocite, covellite). Other important minerals are carbonate (malachite, azurite), oxides (cuprite) and silicates (chrysocolla).

Copper is very similar to zinc, so similar that some of their deficiencies are impossible to differ. Like zinc, copper too exists in soil as ionic or complex form, is immobile, acts as a exchangeable cation and adsorbed strongly by clay minerals (kaolinite) and organic matter. After Pb^{+2}, the Cu^{+2} is strongest adsorbed divalent cations in clay minerals and is the most strongly bound to organic matter than any other micronutrient. The only difference is that Zn appears to be less complexed with organic matter and has sharply reduced availability with increasing soil pH than Cu. Cu^{+2} may substitute for Fe, Mg and Mn in an assortment of minerals like aluminosilicates and carbonates.

Copper concentration reduces with the increase of pH but it increases the adsorption onto mineral colloid surfaces. At pH below 6.9, Cu^{+2} predominates while above pH 6.9 hydrolysis occurs to form insoluble copper hydroxide.

BORON

Boron is the only micronutrient that is non-metal. Its concentration in earth crust is about 10 ppm and plant available boron in soil is about 24 ppm.

Borates and borosilicates are the mineral sources of boron (borax, kernite, colemanite etc.) out of which tourmaline is most common.

Boron is found in soil solution mainly as undissociated boric acid (H_3BO_3) at pH ranging from less than 5 to 9. Plants roots takes the boric acid by the process of diffusion and mass flow. These adsorbed boron constitutes the reserve of B in soil.

Adsorption of B can be increased by increasing fineness of soil texture and quantity of organic matter. Clay minerals like montmorillonite and illite can retain boron.

Increasing soil pH, decreases the availability of boron due to strong boron adsorption to mineral surfaces (fixation). Even boron adsorption by iron and aluminium hydroxides is strongly pH dependent.

Low moisture content in soil can also decrease B mobility in soil and, thus, it cannot be adsorbed by soil roots.

MOLYBDENUM

Lithosphere contains about 2 ppm of Mo and plant availability in soil is about 2.8 ppm.

Mineral sources of Mo is mainly sulphides (Molybdenite) and oxides (Ilsemannite, Wulfenite, Powellite etc.).

Molybdenum is mainly found in soil solution as molybdate (MoO_4^{2-}) that is taken by the plant roots mainly through diffusion. Phosphorus and magnesium enhances the adsorption and translocation of Mo by plants while SO_4^{2-} depresses it.

Mo is the only micronutrient whose concentration increases with increase in pH. MoO_4^{2-} ions are predominant above pH 4.2 and, thus, its concentration increases with liming acid soils unlike other micronutrients.

Like boron, Mo concentration too increases with soil moisture as it increases its mobility for plant adsorption.

Other micronutrients include:

Cobalt which is mainly required by plants for symbiotic nitrogen fixation.
Chlorine is required for photosynthetic reactions in plants.
Silicon gives plants mechanical strength and may help minimize water loss and increase disease resistance. The following Table 16.4 contains additional information of all the nutrients such as its from in the soil, amount available to plants and at which pH they are available in soil for plants to intake.

ESTIMATION METHOD

Amount of nutrients contribution to crops vary for each crop. Amendments (for increasing soil fertility) such as fertilizer or manure are usually required to obtain high yields. Thus, soil testing is used to predict how much and what types of nutrients need to be supplied for a specific field. The main concept of soil testing is that only a proportion of the total quantity of an element in a soil will be available for assimilation of organisms rather than total elemental concentration in soil. This involves extraction of soil samples with a chemical solution followed by a

Table 16.4 Conclusive summary of nutrients and its form, amount of nutrients available to plants and its predominant pH in soil

Nutrient	*Nutrient form in soil*	*Amount available to plants*	*Predominant at pH*
Nitrogen	Organic N, NO_3^-, NH_4^+		–
Phosphorus	$H_2PO_4^-$, HPO_4^{2-}	11–22 mg kg^{-1}	pH < 7.2 ($H_2PO_4^-$) pH > 7.2 (HPO_4^{2-})
Potassium	K^+(solution>exchangeable>fixed>mineral)	5–50 g kg^{-1}	No effect on pH.
Calcium	Ca^{+2}	1.6–35 g kg^{-1}	–
Magnesium	Mg^{+2}	1.5–10 g kg^{-1}	–
Sulphur	SO_4^{2-}	0.9–15 g kg^{-1}	–
Iron	Fe^{+2}, Fe^{+3}	0.36–174 ppm	pH 7.4–8.5 (least predominant)
Manganese	Mn^{+2}	0.60–164 ppm	pH 7
Copper	Cu^{+2}	Trace-32 ppm	pH < 6.9
Zinc	Zn^{+2}	0.8–20.5 ppm	pH < 7.7
Boron	H_3BO_3	Trace-24 ppm	pH 5–9
Molybdenum	MoO_4^{2-}	Trace-2.8 ppm	pH > 4.2

quantification of nutrients in the solution. The chemical extraction is expected to reflect the amount of nutrient that would be available to the crop.

Soil testing is done in the following three steps:

- Sampling and sample preparation
- Analysis of the sample
- Interpretation of the analyzed results

Sampling just prior to planting is recommended for predicting quantity of fertilizer required whereas post-harvest soil sampling is useful for assessing the amount of available nutrients that were not used by the crop.

The availability of elements in soil can be assessed by using chemical extractants which are typically dilute solutions or mixtures of acids, bases, salts and chelates. Chemical analysis assess soil nutrient supplying capacity and other soil properties that affect soil growth like pH, SOM etc. Other than extraction, equilibration, titration and oxidation methods are also used. Biological techniques are usually too expensive and time consuming for routine use.

CONCLUSION

Each nutrient has specific role in plant growth and its absence prohibits the normal growth of the plant. Plant growth rate is decided by both the amount of available and deficient nutrients. Studies of deficiency and toxicity of nutrients in plant growth are important for agriculture and future references. Modern research results helps now grow plants in a controlled environment.

THINK FOR A WHILE

- How does mobility of nutrients affect deficiency in plants?
- Why N_2 cannot be used directly by plants? How does conversion of nitrogen take place?
- Despite being a macronutrient, why phosphorus is found in very low quantity in plants?
- What is the difference between nitrogen and phosphorus fixation?
- Why concentration of sulphur is very high in industrious region?
- Name a micronutrient that is non-metal? Concentration of which micronutrient increases with increase in pH?

REFERENCES

Foth, H.D. (1990). Fundamentals of Soil Science. John Wiley & Sons.

Grundon, N.J. (2006). Nutrient Deficiency and Toxicity Symptoms. *In:* Lal, R., Ed., Encyclopedia of Soil Science. Taylor & Francis. 1144–1146.

Part IV
Soil Biology

Chapter 17
Microbiota

The soil contains billions of organisms representing nearly every phylum. Currently, less than 1% of the soil microorganisms have been cultured, identified and characterized in the soil ecosystem. Soil microorganisms, prokaryotic and eukaryotic, spend all or part of their lives in soil environment. These microorganisms live together by interacting with each other and with other organisms in the soil. Microbial communities are the most complex and diverse group of soil organisms, ranging in size from 0.5 to 5.0 mm, which consists predominantly of bacteria, fungus, actinomycetes and lichens.

VIRUSES

Viruses are the largest number of organisms in soil. They may be more correctly considered as complex molecules of protein and nucleic acid, rather than organisms. They are a group of small non-cellular, infectious, obligate parasites that cannot reproduce by itself and vary in shape, and the virus particle or virion ranges in diameter from 20–250 nm and reach a population of 10^{10} g^{-1}.

Most viruses contain a core of DNA and RNA surrounded by a protein coat (Fig. 17.1). Since they do not contain any metabolic machinery, they are unable to replicate without a living host. Viral infection begins when proteins on the surface of a virion bind to specific receptor proteins on the surface of host cells. Few scientists considered viruses to be a living organisms because of its ability to maintain genetic continuity.

Viruses are often grouped according to the host they infect. But these classifications are not distinct when individual species are able to infect hosts in different domains. These groups are:

- Plant viruses
- Animal viruses

S. Mukherjee, *Current Topics in Soil Science*,
https://doi.org/10.1007/978-3-030-92669-4_17

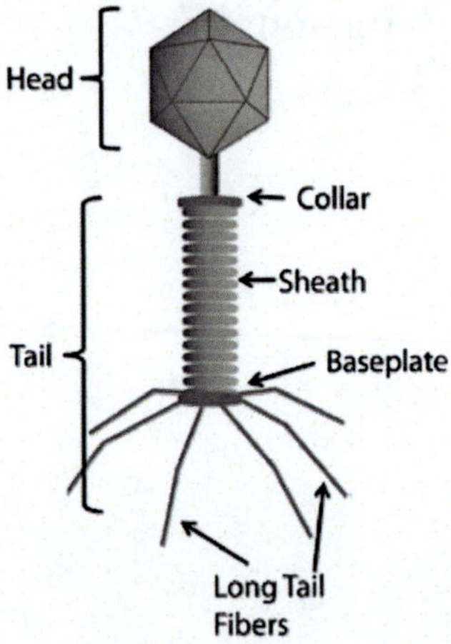

Fig. 17.1 Structural anatomy of a typical soil virus.

- Bacterial or bacteriophage viruses
- Fungal or mycophage viruses

Importance

Since little is known about the vast number of viruses, its importance is also not easy to determine.

- Some viruses of agricultural importance bacteria (*Rhizobium* and *Bradyrhizobium*) have shown to cause a significant decline in bacterial population in soil to the point where crop growth is affected (Sumner, 2000). But they cannot completely eliminate the bacteria as viruses need host to survive. Thus, its population also decreases with the bacteria.
- Since viruses target specific individual, genera or species, they are also used to biocontrol insects.
- Other than suitable hosts, many factors important for survival of viruses in soil are:
- They can survive without a host by adsorbing onto clay particles. Strength of adsorption can be determined by CEC, structure of clay minerals, and electrostatic and hydrophobic interactions.
- Low soil moisture and high temperature can reduce the length of time of survival of viruses.
- Highest number of viruses are found in neutral pH soil.

BACTERIA

Bacteria are the smallest single-celled organisms and exceeds all other organisms by kind and numbers and belong to kingdom Protista and group prokaryotes. Prokaryotes has simple cell structure, i.e. they lack true nucleus and chromosome is found as a closed circular loop within the cytoplasm (Fig. 17.2). Number of bacteria in a soil per gram ranges from 10^8 to 10^9 (Table 17.1). The most common soil bacteria are rod or cocci shaped, a micron (1/25,000 of an inch) or less in diameter, upto a few microns long and approximately10^{-12} g in weight.

Classification of bacteria is extremely difficult because the genetic element in bacteria is mobile and thus genes are flowing continuously from one species to another. Most common classification is grouping according to their reaction with Gram's stain. This depends upon cell wall components; those organisms which retain the stain are termed **Gram-positive** (e.g. *Streptococcus, Listeria*), and those that do not retain the stain are termed **Gram-negative** (e.g. *Escherichia coli, Salmonella*). They can also be grouped as **aerobes** that require a supply of oxygen in soil atmosphere; **anaerobic** can thrive only with an absence of oxygen in the soil atmosphere and **facultative aerobes** that thrive well in an aerobic environment can adapt to an anaerobic environment.

Factors responsible for the bacteria to survive in soil are:

- The concentration of nutrients especially carbon is the most important factor. Bacterial reproduction increases when organic carbon in soil is high. During low concentration of carbon, bacteria returns to low metabolic activity or resting mode.
- pH range for maximum bacterial growth is 4–10 with the optimum being 7.2. At pH<4 excess Al^{+3} and Mn^{+2} becomes toxic whereas at pH>10, concentration of micronutrients decreases due to immobilization and precipitation.

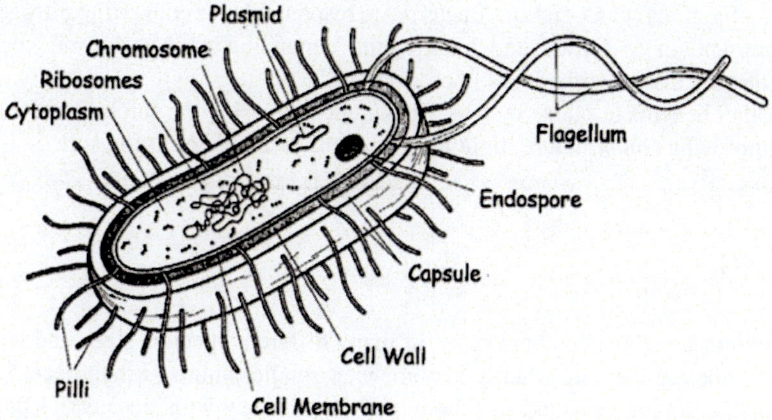

Fig. 17.2 Structural anatomy of a bacterium.

Table 17.1 Relative numbers and biomass of microbial species at 0–15 cm depth of soil

Microorganisms	*Number per gram of soil*	*Biomass (g/m^2)*
Bacteria	10^8–10^9	40–500
Actinomycetes	10^7–10^8	40–500
Fungi	10^5–10^6	100–1500
Algae	10^4–10^5	1–50

Source: Adu and Oades (1978)

- Highest number of bacteria are found under matric potential ~0.01 MPa. Low moisture level can decrease bacterial growth due to desiccation and accumulation of toxic products. High moisture level can decrease bacterial activity due to decrease in O_2 level by reduction in gas exchange between atmosphere and soil.
- Many contaminants in soil can decrease bacterial population, excess Zn and Cd in soil can decrease bacterial population by 75%.
- Since soil has a wide range of temperature, both diurnally and seasonally, bacteria too can survive for wide range in temperature (–10 to 110 °C).

Importance

- Some species of bacteria release nitrogen, sulphur, phosphorus and trace elements from organic matter. Others break down soil minerals, releasing potassium, phosphorus, magnesium, calcium and iron.
- Other species make and release plant growth hormones, which stimulate root growth.
- Several species of bacteria transform nitrogen from a gas in the air to forms available for plant use, and from these forms back to a gas again. A few species of bacteria fix nitrogen in the roots of legumes, while others fix nitrogen independently of plant association. Bacteria are responsible for converting nitrogen from ammonium to nitrate and back again, depending on certain soil conditions (already discussed in detail in chapter 16).
- Other benefits to plants by bacteria include increasing the solubility of nutrients, improving soil structure, fighting root diseases and detoxifying soil.

FUNGI

Fungi are heterotrophs that comes in many different species, sizes and shapes in soil. Although they are usually in somewhat smaller numbers than bacteria, fungi dominate the biomass and metabolic activity in many soils because of their relatively large size and branching. They dominate about 10–30% of soil rhizosphere.

Fungi typically grow or germinate from spores and form a threadlike structure, called mycelium which are composed of individual hyphae, 0.5–10 μm in diameter (Fig. 17.3). Mycelium is the working structure that absorbs nutrients, continues to grow and eventually produces reproductive structures that contain spores.

In contrast with bacteria, which are prokaryotic, unicellular and is limited to surface erosion in place, fungi are eukaryotic and generally produce filamentous hyphae that can penetrate and explore microhabitats of the soil to extract new substrate and proliferate both within and between dead plant cells. Even in many ecosystems, fungi are the most abundant primary decomposers than bacteria. This is not only because of their hyphal network but also because they can produce a range of enzymes such as lignin that are capable of degrading complex cell wall components, enabling them to gain access to more labile compounds that occur within the cell. Another advantage of fungi over bacteria is their ability to transport nutrients through their hyphal network to zones of exploitation, i.e. they have the capacity to exploit new substrates, even when nutrients such as N and P are limiting (Boddy, 1999).

Importance

- Many fungi aid plants by breaking down organic matter or by releasing nutrients from soil minerals.

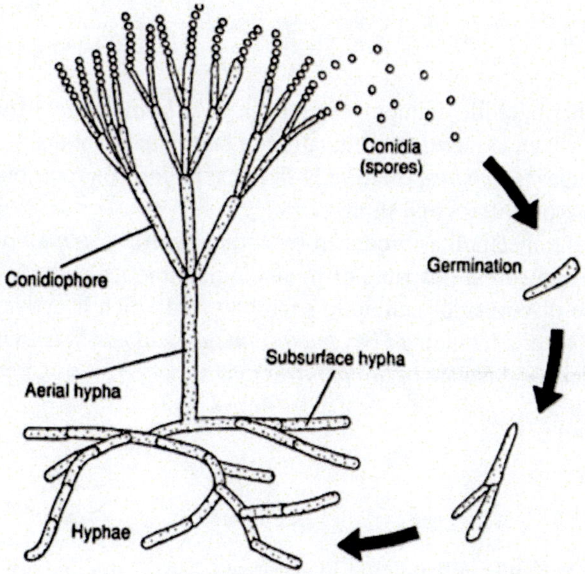

Fig. 17.3 Microscopic structures of a septate fungus showing asexually produced conidia that leave the fungus and germinate to produce a new mycelium.

- Some fungi produce plant hormones, while others produce antibiotics including penicillin (e.g. *Penicillium chrysogenum, Penicillium nalgiovense*).
- Their tolerance for acidity makes them particularly important in acidic forest soils.
- They act as plant pathogens (*Candida albicans)*, they bind soil particles together, thereby enhancing soil structural stability, and they provide a food source to microbial-feeding fauna.

ACTINOMYCETES

Actinomycetes are prokaryotic organisms that resembles fungi in that they produce a branched filamentous network (less extensive than fungi), the production of extracellular enzymes and absorptive nutrition. Although actinomycetes were previously considered as fungi but now as bacteria because they have a very simple cell structure and are about the same size in cross-section. Some of the actinomycetes reproduce by spores like bacterial cells. They are gram-positive bacteria.

There are no definite estimate over how many species of actinomycetes are there but it is considered over 400. The numbers of actinomycetes may vary from 1 to 36 million/g in soil. Although actinomycetes develop best in moist, warm, well-aerated soil, they are functionally important in arid-region, salt-affected soils (Islam and Wright, 2006).

Importance

- Certain thermophilic actinomycetes (*Agaricus brunnescens)* forms composted substrate which are used for cultivation of button mushrooms.
- Some species (*Frankia* spp.) form N-fixing symbiosis on roots of a wide range of perennial woody trees and shrubs.
- Some of the metabolites produced by actinomycetes (*Streptomyces* spp.) when isolated from soil are capable of producing antibiotics under laboratory conditions. The discovery of antibiotic streptomycin, which is active against a wide range of bacteria including *Mycobacterium tuberculosis*, was found on work carried out on a soil isolate of *Streptomyces griseus.*

ALGAE

Algae are eukaryotic cells ranging in size from 2 to 20 μm and are capable of producing their own food with the help of photosynthesis. Several hundred algal species have been identified in soil, but only a small number of species are prominent throughout the world. Most species of algae live in the upper half-inch of the soil but may also be found at depths down to 15–20 cm below the surface, where there is insufficient light for photosynthesis.

Cyanobacteria (or blue green algae) is responsible for aggregating soil particles to form biological soil crusts especially in conditions where water and nutrients are limited. This is done by physical binding of soil particles by entangled filaments, adhesion to mucilaginous sheaths or slime layers excreted by cyanobacterial trichomes and attachment of particles to sites along the cyanobacterial cell walls (Bar-Or and Danin, 1989). This binding increases soil stability, organic matter content of the crust preventing soil erosion. *Chlamydomonas* and *Asterococcus* species are known to excrete polysaccharides (gelationous material) which adhere to soil particles into single consolidated layer improving the soil structure.

Some species of blue-greens algae like *Tolypothrix tenuis* can fix their own nitrogen and their importance in improving soil fertility for sustainable agriculture in submerged and irrigated rice cultivation is well recognized. Although algae grow best under moist to wet conditions, but some thrive in hot or cold environments too.

CONCLUSION

The soil is a complex and heterogenous component of the terrestrial ecosystem with a diverse community of microbes dominated by viruses, bacteria, fungi, actinomycetes and algae. The microbial community is vital to the cycle of life on earth and some of their ecological functions include litter decomposition, nutrient cycling, bioremediation, C sequestration, and improvement of the physical properties of the soil. With the diversity in a soil ecosystem, many types of community interactions are reported including parasitism, competition and symbiosis.

THINK FOR A WHILE

- Why are viruses used in biocontrolling insects in agriculture?
- How does moisture level affect bacterial population?
- Why fungi are the most abundant primary decomposers than bacteria in some ecosystems?
- Why actinomycetes were previously considered a fungi?
- How does algae affect soil structure? Why it is used as an biofertilizer?

REFERENCES

Adu, J.K. and Oades, J.M. (1978). Physical factors influencing decomposition of organic materials in soil aggregates. *Soil Biology and Biochemistry.* **10:** 109–115.

Bar-Or Y. and Danin, A. (1989). Mechanisms of Cyanobacterial crust formation and soil stabilization in the northwestern Nagev. *Proc. Ier. Soc. Microbiol., Haifa*, 13.

Boddy, L. (1999). Saprotrophic Cord-Forming Fungi: Meeting the Challenge of Heterogeneous Environments. *Mycologia,* **91:** 13–32.

Islam, K.R. and Wright, S.R. (2006). Microbial Communities. Taylor & Francis.

Sumner, M.E. (2000). Handbook of Soil Science. CRC Press.

Chapter 18
Soil Fauna

Soil fauna consists of large number of animals that have significant effects on structural and functional properties of soil. They are classified into microfauna, mesofauna and macrofauna (Fig. 18.1) and are not only classified based on body size but also method of sampling. Their functions in the soil are also different. For example, microfauna are rarely involved in litter comminution (reducing size), the mesofauna do attack plant litter, but their overall contribution is small (termites are exception to this) and the macrofauna have the major effect on litter decomposition (Wood, 1995).

Castro-Huerta et al. (2015) studied the roles of macrofauna, mesofauna and microfauna on litter decomposition, and the way they were modulated by soil use. This was done by keeping litter bags of different mess sizes in different ecosystems (naturalized grasslands, recent agriculture and intensive agriculture fields) and then calculated the decomposition rate (Olsen, 1963) as:

$$RM = IM \times e^{-kt}$$

where RM is the remaining mass, IM the initial mass, t the time (days), and k the decomposition rate.

The study showed that in the naturalized grasslands, the macrofauna had the highest contribution to decomposition whereas in agricultural system it was the combined activity of the macro- and mesofauna. In the intensive agricultural use it was the mesofauna activity that has the highest contribution to decomposition. Decomposition rate of microfauna was significantly higher in less disturbed naturalized grassland system when compared to the two agricultural systems. This suggests that microfauna is the most sensitive group to anthropic disturbances, and thus should be particularly taken into account when devising sustainable agricultural practices (Fig. 18.2).

S. Mukherjee, *Current Topics in Soil Science*,
https://doi.org/10.1007/978-3-030-92669-4_18

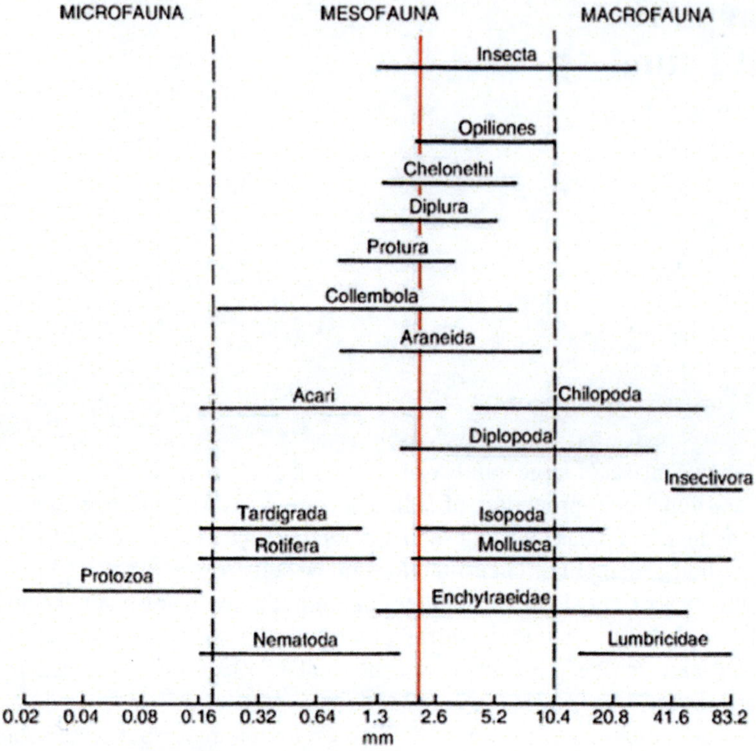

Fig. 18.1 Size classes in soil fauna groups, according to Wallwork (1970) (red line shows 2 mm boundary).

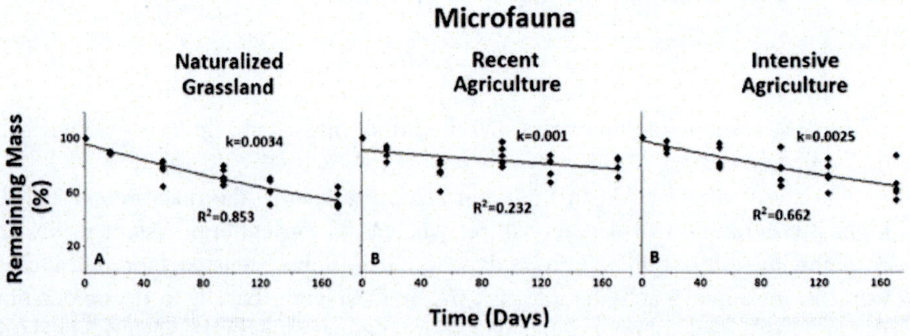

Fig. 18.2 Graph of decomposition rate (k) of microfauna in different ecosystems.

PROTOZOA

Protozoa are free living microfauna in soil that feed on microbes and are very important organisms in nutrient cycling. They mainly live in water films, covering soil aggregates and organic matter. Due to their short life span, they can adjust quickly to environmental changes. Protozoa have greater assimilation efficiency and turnover time than any other soil fauna.

Protozoa has lower concentration of nitrogen in their cells than the bacteria they prey on. The excess nitrogen is excreted from their system as NH_4^+ which are taken by plants and other microorganisms. Since protozoa prey mostly on bacteria, it helps their population in controlling which also influences the decomposition rate and soil aggregation process.

They belong to two phyla: Sarcomastigophora and Ciliophora and four ecological groups: flagellates, naked amoebae, testacea and ciliates (Fig. 18.3):

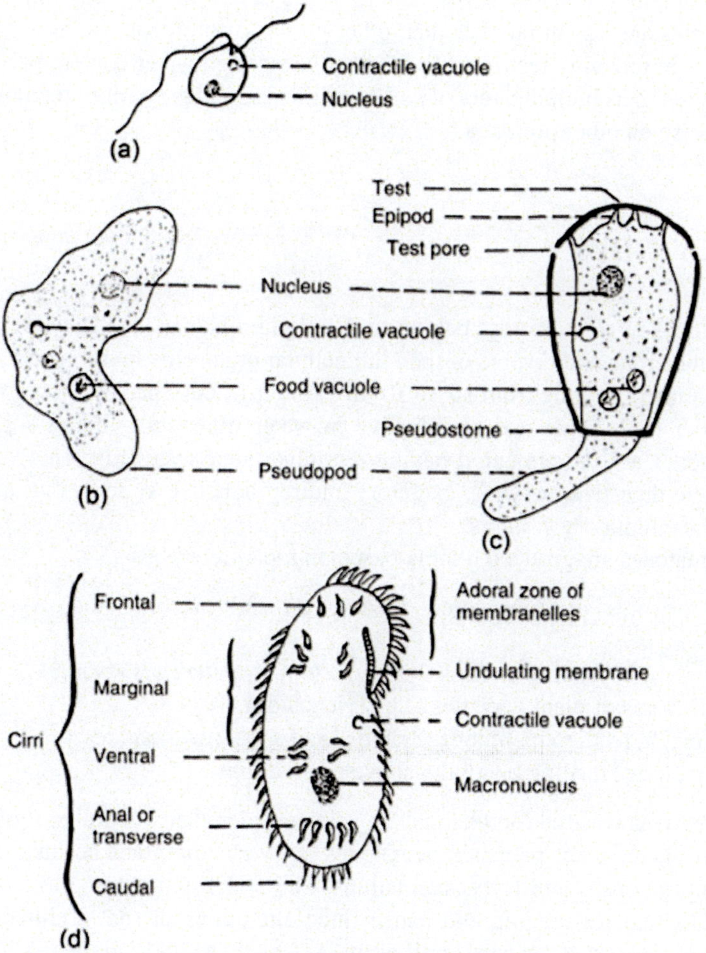

Fig. 18.3 Morphology of four types of soil protozoa: (a) flagellate (*Bodo*); (b) naked amoeba (*Naegleria*); (c) testacean (*Hyalosphenia*) and (d) ciliate (*Oxytricha*).

- The most numerous and active protozoa are flagellates which plays a significant role in nutrient turnover with bacteria as their principal prey. Many flagellates are saprozoic, i.e. feeding on dissolved organic nutrients.
- Naked amoebae are among the more voracious of soil protozoa feeding on bacteria, fungi, algae, and other fine particulate organic matter. They are considered more suited to soil because their sliding motion on surfaces enables them to feed on soil particles and roots, and their flexible cells are adapted to feeding within thin water films around soil particles that would otherwise be considered inaccessible to predators.
- Testate amoebae are often less numerous than the naked amoebae, except in moist, forested systems. They differ from naked amoebae in the presence of a test that partially encloses the cell, with an aperture from which the pseudopodia emerge. This provides the amoeba with shelter from predators and environmental conditions and also helps them to be preserved for very long time.
- Ciliates are less in number than other groups of protozoa. Being larger in size, they are probably restricted to feeding during high water content periods. Thus, they serve as bioindicators of soil conditions. Ciliates not only feed on bacterias but also on other protozoa.

NEMATODES

Nematodes or roundworms belong to the phylum Nematoda. They are one of the most numerous and diverse of the multicellular organisms found in any ecosystem with numbers ranging from 10^6 to 10^7 m^{-2}. About 15,000 species known vary in size from 0.5 to 2.0 mm and mainly live in water films surrounding soil particles. Nematodes are unsegmented having wormlike apperance (Fig. 18.4) and have a complete digestive, nervous, excretory and reproductive systems but lack respiratory and circulatory systems.

Nematodes are grouped mainly based on food preferences:

- Bacterivores or microvores, feeding on soil bacteria (e.g. *Agrobacterium tumefaciens*)
- Fungivores, feeding on fungal mycelia (e.g. *Aphelenchus avenae*)
- Herbivores or plant parasites (e.g. Tylenchida)
- Predators feeding on other nematodes and small invertebrates
- Omnivores, feeding on all the above given things.

Due to their varied food habits, they are considered an integral part of food web. Among them, plant parasites nematodes are very important because it damages agricultural crops and forest plantations. The effect nematodes have on plants is generally species-specific and can include alterations in root architecture, water transport, plant metabolism, or all of these (Coleman and Wall, 2015). They generally have cell wall degrading enzymes that modify plant cell walls and cell metabolism. Nematodes are very serious problem for pineapple production in Hawaii,

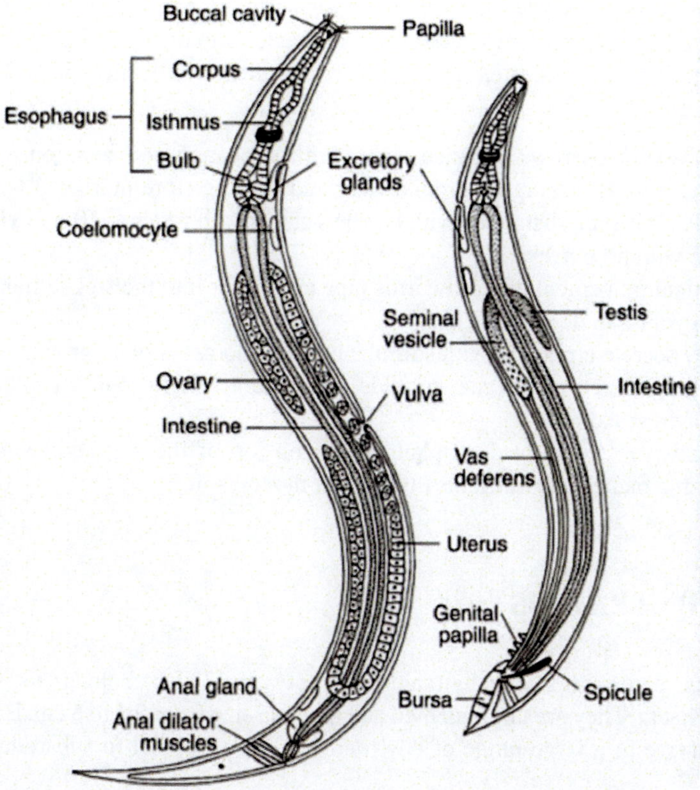

Fig. 18.4 Anatomy of a typical nematode showing major structures.

where the soil for a new planting is routinely fumigated to reduce nematode population (Foth, 1990).

EARTHWORMS

Earthworms are the most commonly known animals in soil and has the most significant effects on soil structure and function. They belong to the phylum Annelida and class Oligochaeta. They are most abundant in forests and grasslands of temperate and tropical regions and their population ranges from 1 to 850 per m^2. Tillage decreases earthworm population density.

Earthworms prefer a moist environment with an abundance of organic matter and calcium supply in soil. As a result, they tend to be least abundant in dry and acidic sandy soils and neither prefer too hot climate nor too cold. For example, *Millsonia anomala* becomes inactive when the soil moisture content is less than 7% and

Allolobophora caliginosa is found at the soil surface when the temperature is greater than 5°C but at lower temperatures moves deeper into the soil (Fig. 18.5).

As mentioned earlier earthworms exert significant influences in soil processes. They are:

- Earthworm burrows enhance water infiltration and soil aeration by reducing water runoff, recharging groundwater and helping store more soil water for dry spells. It is seen that fields with burrows can absorb water 4–10 times higher than field with no burrows.
- Vertical movement of earthworms pipe air deeper into the soil stimulating nutrient cycling in deeper levels.
- They secrete a plant growth stimulant which increases plant growth.
- Earthworms tunnels can provide deep root penetration of plants during drought season.
- Earthworms feed on dead plant matter on top of the soil and redistribute the organic matter and nutrients throughout the top soil.

ENCHYTRAEIDS

Besides earthworms, enchytraeids or potworms also belong to the family Oligochaeta. They are unpigmented and range in size from 0.1 to 5 cm. Enchytraeids can survive in a large range of environment, from tropical to subarctic. They can

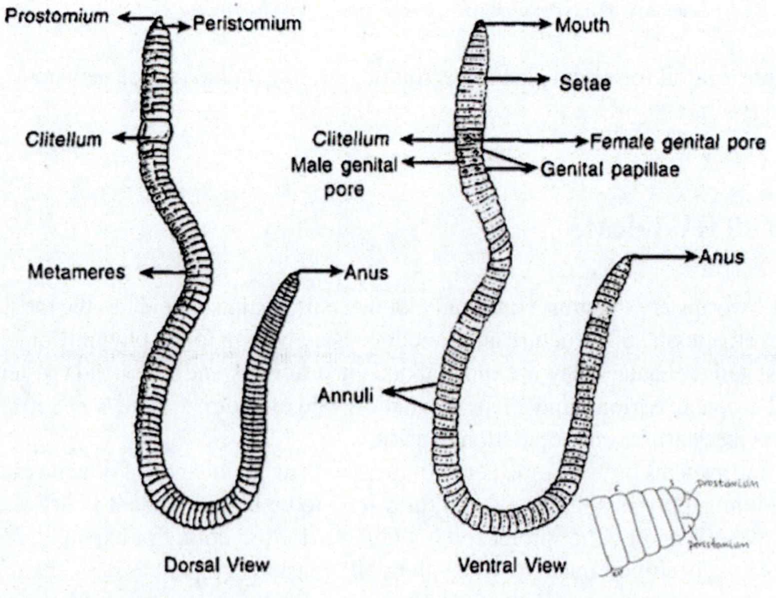

Fig. 18.5 Morphology of earthworm structures with dorsal and ventral sections.

feed on both mineral and organic matter and a large part of their diet mainly includes finely divided plant materials, often enriched with fungal hyphae and bacteria.

Enchytraeids are very important to soil processes:

- Enchytraeids affect soil structure by producing fecal pellets, which, depending on the animal size distribution, may enhance aggregate stability in the 600–1000 μm aggregate size fraction. In forest floors, these pellets are composed mainly of fine humus particles, but in mineral soils, organic matter and mineral particles may be mixed into fecal pellets with a loamy texture (Coleman and Wall, 2015).
- Enchytraeids also enhances the decomposition of litter and nutrient mineralization by affecting soil microorganisms. For example, grazing activities of *C. sphagnetorum* increased the hyphal length and respiration of the fungus *Mortierella isabellina* thus increasing organic matter turnover, microbial metabolic activity and accelerate the release of nutrients.

ARTHOPODS

Arthopods are mainly divided into micro- and macroarthopods depending on their size and method of sampling. Microarthopods typically range in size from 0.2 to 10 mm where macroarthopods ranges from 10 mm to even 15 cm (for centipedes). Microarthopods are sampled by collecting a fragment of habitat whereas macroarthopods are mainly sampled by hand sorting with various other methods. Also the major difference between them is their impact in soil, i.e. macroarthopods are capable of restructuring soil profiles or relocating large amounts of soil whereas microarthopods must take advantage of existing pore spaces (Coleman and Crossley, 1996).

Microarthopods are dominated mainly by two groups, viz. mites (Acari) and springtails (Collembola). Many microarthropods feed on fungi and nematodes, thereby linking the microfauna and microbes with the mesofauna (Chamberlain et al., 2006). Environment with large organic matter content support high numbers (200,000 m^2) of microarthropods like temperate and coniferous forests. Tropical forests, where the organic layer is thin, contain fewer microarthropods.

Macroarthopods are dominated by ants, termites, spiders, beetles etc. They mainly feed on microarthopods and often fall prey to larger vertibrates thus creating a link between microfauna and macrofauna. For example, Collembola are important food items for spiders, especially immature stadia whereas other macroarthropods, such as cicadas, emerging from soil may serve as prey for some vertebrate animals.

Arthopods are mainly decomposers. Their role is to eat and shred the large particles of plant and animal residues and bury it, bringing into contact with other soil organisms that further decompose it.

CONCLUSION

Soil fauna contribute significantly to the structure, texture, and nutrient cycling of soil through bioturbation and interaction with microbiota or other faunas. Tillage destroys soil fauna habitat and food, hence the population is usually higher in undisturbed soil than in tilled soil. Soil fauna exists in food webs containing several trophic levels. They act as both prey and predator thus connecting microfauna to macrofauna making the structure of food web much more complicated with few missing links.

THINK FOR A WHILE

- How does the functions change with size of soil fauna?
- What type of protozoa are most suited to soil and why?
- What type of protozoa serve as bioindicators?
- How does agriculture get affected by parasitic nematodes?
- How does earthworm and enchytraeids affect the structure of soil?
- What organisms forms a link between macro- and microfauna and how?

REFERENCES

Castro-Huerta, R.A., Falco, L.B., Sandler, R.V. and Coviella, C.E. (2015). Differential Contribution of Soil Biota Groups to Plant Litter Decomposition as Mediated by Soil Use. *Peer J.*, **3:** e826.

Chamberlain, P.M., Bull, I.D., Black, H.I.J., Ineson, P. and Evershed, R.P. (2006). Collembolan trophic preferences determined using fatty acid distributions and compound-specific stable carbon isotope values. *Soil Biol. Biochem.* **38:** 1275–1281.

Coleman, D.C. and Crossley, D.A. (1996). Fundamentals of Soil Ecology. Academic Press, New York, USA.

Coleman, D.C. and Wall, D.H. (2015). Soil Fauna: Occurrence, Biodiversity, and Roles in Ecosystem Function. Elsevier Inc.

Foth, H.D. (1990). Fundamentals of Soil Science. John Wiley & Sons.

Olsen, J.S. (1963). Energy Storage and the Balance of Producers and Decomposers in Ecological Systems. *Ecological Society of America.* **44:** 322–331.

Wallwork, J.A. (1970). Ecology of Soil Animals. McGraw-Hill, London.

Wood, M. (1995). Environmental Soil Biology. Chapman & Hall.

Chapter 19
Organism Interactions with Soil Processes

In previous chapters, importance of different types of microbiota and soil fauna were introduced. This chapter mainly concerns with functional significance of these organisms and their interactions with one another for ecosystem processes, especially decomposition and nutrient cycling.

The two main sources of organic matter in soil is the death of roots and decomposition of litter. The litter at the surface of the soil can be brought down by the soil fauna which later decompose by the microbes or it can be dissolved or dispersed. Decomposition is very important for nutrient cycling. Most nitrogen and a portion of the phosphorus and sulphur required for plant growth are supplied through decomposition of organic matter, relying on the activities of soil biota. Soil fauna too influences nutrient cycling by feeding on microbes and other fauna. Since nutrient cycling has been discussed in detail in chapter 16, this chapter only deals with organism's role in the process and their interaction.

MICROBIAL TRANSFORMATION

Microbes acquire energy by oxidation and reduction reactions that are also important for cell growth and replication. These elements are assimilated by the microbes which utilize them by producing cellular components like proteins and phospholipids. During metabolism, some byproducts are discarded by the microbes that can alter the surrounding environment. They sometimes also produce enzymes that increases bioavailability of certain elements. Mineralization is the main function of these microbes by which they convert immobilized elements into inorganic form that can be easily taken by the plants. Potassium, sulphur and some metals are mainly transformed by microbes that either increases the bioavailibility or becomes toxic to plants.

S. Mukherjee, *Current Topics in Soil Science*,
https://doi.org/10.1007/978-3-030-92669-4_19

NITROGEN

Dead organic matter contain most of the nitrogen in the form of proteins, nucleic acids and chitin. These polymers are too large to pass through microbial membranes, so microbes produce extracellular enzymes like proteinases, ribonucleases, and chitinases that break them down into smaller, water soluble subunits (Dissolved Organic Nitrogen) that can be absorbed by microbial cells for their growth. Absorption of C or N from Dissolved Organic Nitrogen (DON) mainly depends on what limitations does microbes have. If microbes are C limited, they take up C from DON and excrete NH_4^+ that can be taken up by plants (N mineralization). Whereas when N is limited, they take up inorganic N from soil and limit the plant nutrition (immobilization). Thus, both processes take place simultaneously and the balance depends onC : N ratio of the organic substrate.

Microbes mainly take up N in the form of NH_4^+, although, some bacteria can also assimilate NO_3^- in their system. Nitrogen fixation is mainly carried out by organisms like bacterium *Azotobacter* sp., the obligate anaerobe *Clostridium* sp., the photosynthetic blue green bacterium *Anabaena* sp., and the symbiotic bacteria *Rhizobium* sp. and *Frankia* sp. Most of these species live in a symbiotic relationship with plant (exception Frankia), forming nodules in the root where they receive carbohydrate from the plant to meet their energy needs and in turn they supply the plant with amino acids formed from reduced N.

Some free living bacterias like *Nostoc* sp. too can fix nitrogen by decomposing litter and soil organic matter. Although they fix relatively small amount but they have significance mainly in extreme environment where there is sparse cover of vascular plants.

PHOSPHORUS

Plants mainly take P in the form of inorganic phosphate which is mainly found in two forms:

- Labile P pool, mainly found in minerals that can be easily absorbed by plants.
- Occluded P, which precipitates with Fe and Al compounds, and becomes insoluble and unavailable to plants.

Organic P which constitute about 29–60% of P in soil comes from inositol phosphates but their role to P cycling is not well-known.

Since most of the P comes from minerals, so most of the transformations are geochemical rather than biological. Despite, soil microbes play an important role in P cycling, i.e. mineralization. Some microbes contain enzymes such as phosphatases which act by cleaving ester bonds in organic matter to liberate phosphate (PO_4^3).

Microbes contain around 20–30% of the total soil organic P pool which is much larger than the proportion of C or N held in microbes. Since plants also needs P, microbes compete aggressively with plants for available P in soil. The mineralization of organic P is partly regulated by the C : P ratio of substrates, i.e. when C : P ratio rises above 100, then P is immobilized by microbes.

MYCORRHIZAE

Mycorrhizae (also called fungus roots), are symbiotic association between plant roots and soil fungi. Hyphae of the fungus penetrate deep into the roots and grow mycelium both on exterior and interior of the root. The plant provide the fungus with C and in return fungus provides the plant with inorganic nutrient in the form of N and P. The hyphae becomes extension of the root and act as root hair to absorb nutrients. Continued growth of the hyphae outside the root results in the formation of a mantle that completely surrounds the root which protects the plant from soil born pathogens (Fig. 19.1). It also helps the plant as resistant to heat and drought. Mycorrhizal fungi is also known to be one of the largest C sinks for primary productivity.

Mycorrhizal fungi play very important role in P uptake by plants. Fungi with the help of hyphae penetrate deep into the soil accessing P that is beyond the roots. This is done by producing phosphatase enzymes that releases phosphate ions for plants. Second, they produce low molecular weight organic acids, such as oxalates, which enhance the availability of soil P by increasing weathering rates of P contained in clay minerals, and by complexing cations (e.g. Ca, Fe and Al) that would otherwise bind to phosphates, thereby taking them out of soil solution (Bardgett, 2005).

Mycorrhizae is also very important for nitrogen fixation. It not only enhances uptake of ammonium from soil, but also has the capacity to uptake dissolved organic nitrogen in the form of amino acids and proteins. This is because it produces carboxyl-proteinase enzymes that have the ability to cleave proteins into constituent

Fig. 19.1 Microscopic view of fungal hyphae (hair like) surrounding a plant root in soil (Lambers et al., 2008).

amino acids. They can also transport fixed N from legumes, through soil, to neighbouring plants with the help of hyphal networks.

SULPHUR

Organic S mainly found in organic matter are decomposed by microbes such as autotrophic bacteria (*Thiobacillus* sp.), fungi (*Penicillium* sp.) etc. They transform organic S to soil organic sulphate or C bonded S which is then mineralized to soluble sulphate taken by plants. This is mainly done with the help of some enzymes such as sulfatases that hydrolyse sulphate esters. In-turn, it can also be immobilized back into organic S by reduction. Most of the S immobilized by microorganisms resides in amino acids. This is called the sulphur cycle. The reduction of SO_4^{2-} to H_2S is carried out by anerobic bacteria such as *Desulfovibrio desulfuricans* which is toxic to most aerobic organisms including crop plants, and may cause precipitation of metal sulphides and hence the characteristic black colour of anaerobic muds (Fig. 19.2).

MICROBE-FAUNA INTERACTION

As discussed before, the main function of microbes in nutrient cycling is mineralization. But these mineralization by microbes can be affected by soil fauna. The main effects are:

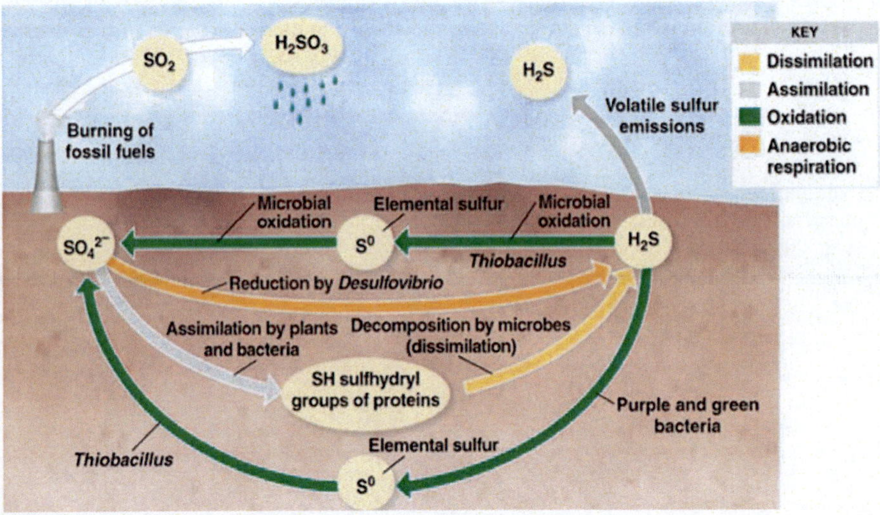

Fig. 19.2 Role of microbes in sulphur cycle.

- Selective feeding of microbes by soil animals which alters microbial activity, abundance, and community structure.
- Soil fauna alters the form of the organic matter in soil by fragmentation and mix organic matter inputs to soil, increasing its susceptibility to microbial attack.
- There are some soil fauna such as protozoa that has non-nutritional effect on plant growth.

SELECTIVE FEEDING

Selective feeding of soil fauna is not unknown to biologist. Most of the effects on selective feeding, nutrient mineralization and primary production are considered positive. Increased turnover rate, activity, and respiration of grazed microbial population increases C mineralization whereas animal excretion N mineralization. This is due to the fact that soil fauna has much lower assimilation efficiencies than microbes and thus they excrete nutrients, not required by them thereby increasing its bioavailability. For example, protozoa preying on bacterial population are assumed to release about one-third of the nitrogen consumed (Fig. 19.3).

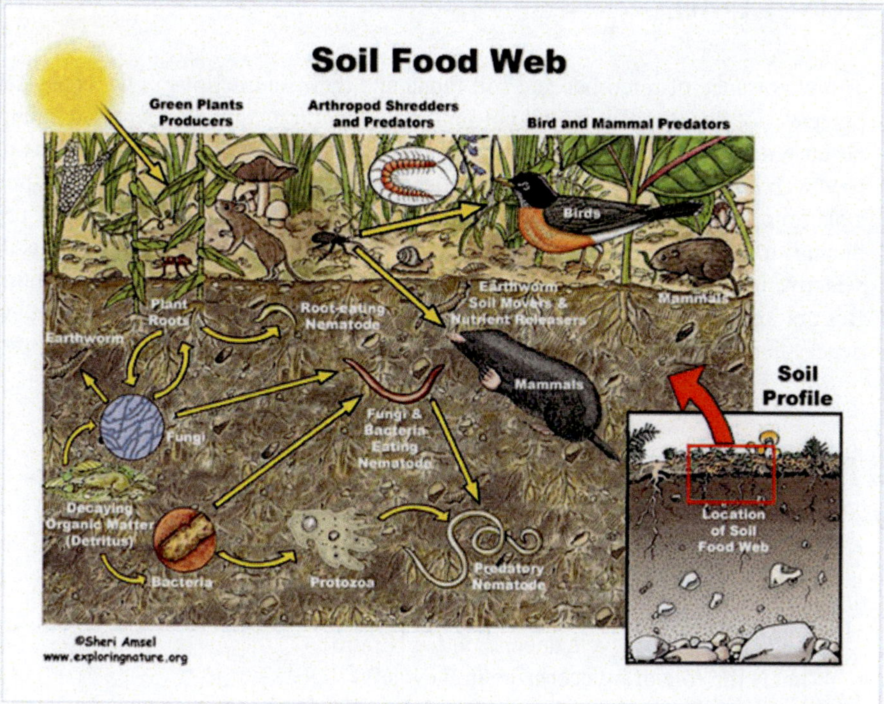

Fig. 19.3 Interactions between organisms of different trophic levels and subsequent food web.

It is also evident that interaction between decomposer organisms can affect performance of higher trophic groups that benefit from enhanced plant nutrition. For example, plant sucking aphids performed better and increase in biomass where the host plant was grown in the presence of microbial-feeding Collembola or earthworms thereby increasing N turnover.

NON-NUTRITIONAL EFFECT

Not all effect of interactions between soil fauna and microbes are positive. It highly depends on resource availability. For example it is seen that in N-limited soils, nutrients released by fauna grazing on fungi are often rapidly re-utilized by microbes, and hence do not become available for plant uptake (Visser et al., 1981; Bardgett et al., 1993; Cole et al., 2002). Also there are some protozoa that indirectly influence root architecture by influencing bacterial growth. These bacterias produce a plant growth-promoting hormones such as auxins. Thus, selective grazing by protozoa in the rhizosphere significantly stimulates the growth of auxin producing bacteria which directly influences the root growth of plants.

CONCLUSION

The importance of microbes and soil fauna and their interactions in the ecosystem processes of decomposition and nutrient mineralization cannot be stressed enough because it determines the availability of nutrients in plants that directly affects the food web. Main function of microbes is decomposition whereas soil fauna is responsible for altering the biophysical properties of soil by litter transformation and producing structures for microbial habitat and material transfer. Most significant is the selective feeding of microbes by soil fauna that directly influences the nutrient flux. But not all effects of soil fauna in nutrient cycle is positive. There are ample evidences that suggest soil fauna can also cause non-nutritional effect on plants.

THINK FOR A WHILE

- How do microbes acquire energy? How do they increase bioavailability of certain elements?
- What is dissolved organic nitrogen?
- How does nitrogen cycle depend on C : N ratio?
- What is the role of myccorhizae in P cycle?
- Why anaerobic muds are usually black in colour?
- How does soil fauna increase C and N mineralization?
- How does protozoa influence root architecture of plants?

REFERENCES

Bardgett, R.D. (2005). The Biology of Soil: A Community and Ecosystem Approach. Oxford Press.

Bardgett, R.D., Frankland, J.C. and Whittaker, J.B. (1993). The effects of agricultural management on the soil biota of some upland grasslands, *Agriculture, Ecosystems & Environment.* **45:** 25–45.

Cole, L. and Bardgett, R.D. (2002). Soil animals, microbial interactions and nutrient cycling. *In:* Lal, R. (Ed.), Encyclopedia of Soil Science. Marcel Dekker, New York, pp. 72–75.

Lambers, H., Raven, J.A., Shaver, G.R. and Smith, S.E. (2008). Plant nutrient-acquisition strategies change with soil age. *Trends in ecology & evolution.* 23: 95–103.

Visser, S., Whittaker, J.B. and Parkinson, D. (1981). Effects of collembolan grazing on nutrient release and respiration of a leaf litter inhabiting fungus. *Soil Biol Biochem,* 13.

Chapter 20
Impact of Pesticides and Fertilizers on Soil Microbial Diversity

Pesticides are chemical or biological substances or a mixture of substances intended for repelling, destroying or controlling any pest, or regulating plant growth. Most common pesticides are herbicides (kill weeds or other plants), insecticides (kill insects and other arthropods) and fungicides (kill or prevent the growth of fungi and their spores).

With increasing demand, use of pesticides are increasing globally (Fig. 20.1). For the approval of any pesticides, its effect on microbial diversity and soil fertility is required. Some pesticides enhances the growth of soil microbes whereas others have a negative or no impact. From previous chapters in Soil Biology, it is well known that soil microorganisms play a very important role in soil fertility and productivity. Degradation of organic matter, nutrients recycling, humus formation, soil structural stability, nitrogen fixation, disease biocontrol and biochemical transformation are few of the many role of microorganisms in soil. Thus, it is important to know the influence of pesticides on microbial diversity.

Similarly, soil fertilizers too can affect soil microbial communities. Its effect depend on the pH of the soil. It is seen that microbial population decreases in soil with pH less than 5. Duration of application of fertilizers is also an important factor for determining the population change of microbes. Measurement of microbial diversity is done by measuring the soil respiration or CO_2 production in soil.

EFFECT OF PESTICIDES

Depending on the chemical nature of pesticides, it undergoes many degradative, transport and adsorption/desorption processes. Its effect on soil microorganisms are controlled by many environmental factors such as (Table 20.1):

- bioavailability
- bioactivity

S. Mukherjee, *Current Topics in Soil Science*,
https://doi.org/10.1007/978-3-030-92669-4_20

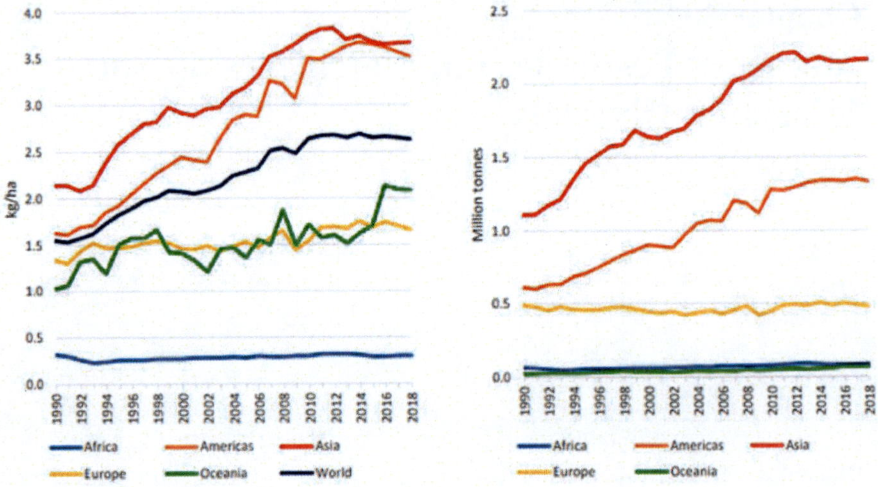

Fig. 20.1 Regional pesticides use per cropland area and absolute amounts by continents (FAOSTAT, 2020).

Table 20.1 Effect of different pesticides on microbial diversity

Pesticides	*Soil source*	*Result*
Butachlor (H)	Rice growing tracts	Reduces the population of *Azospirillum* and anerobic nitrogen fixers in a non-flooded alluvial soil at pH 6.2
	Paddy soil	Stimulates the population of anaerobic fermentative and sulfate-reducing bacteria while inhibites acetogenic bacteria
Carbofuran (I)	Rice growing tracts	Stimulates the population of *Azospirillum* and other anaerobic nitrogen fixers in the soil
Malathion (I)	All soils	Increases the population of denitrifying bacteria
Chlorpyrifos (I)	Soybeans, corn	No significant change in population of bacteria
Glyphosate (H)	Ponderosa pine plantation	Produces a non-specific, short-term stimulation of bacteria at a high concentration
Iprodione (F)	Strawberry orchard	Increases the population of soil bacterial community
Captan (F)	Apples and pears	Decreases the population of *Rhizobium ciceri*
Isoproturon (H)	Farm soil	Increases the population of bacteria *Sphingomonas* spp.
Fenpropimorph (F)	Researched Plot	Inhibited the growth of active fungi and bacteria in soil

H – Herbicide Adapted from Lo (2010) and Hussain et al. (2009).
I – Insectiside
F – Fungicide

- persistence
- concentration
- toxicity of pesticides

Soil factors are also important to understand the relationship between pesticides and microbes. They are:

- texture
- vegetation
- tillage system
- organic matter

For example, application of pesticides in conventional tillage system tends to affect more negatively than no-tillage system. Thus, population of mycorrhizal colonization decrease after 12 days of application of herbicide fomesafen. Also application of herbicides tends to be more severe in coarse textured soil. Menon et al. (2004) reported more inhibitory effect of chlorpyrifos and quinalphos in the loamy sand than in the sandy loam soil. This is due to greater bioavailability (less sorption) of the pesticides in loamy sand, resulting from low clay content and organic carbon.

Toxicity of pesticides have severe effects on many microbial community. Persistence or duration of pesticides can also have similar effects. It is seen that repeated applications of atrazine can significantly reduce the intensity of soil microbes.

Combination of pesticides with other fertilizers or heavy metals have adverse affects on soil microorganisms than using a pesticide alone. For example, butachlor when applied in combination with cadmium, can reduce microbe count in soil.

After application, pesticides undergoes numerous biochemical and physiochemical transformation to form other metabolites which can be more toxic to soil microbes. Virag et al. (2007) studied the effects of pesticides and their degradative products on numerous microorganisms and found that herbicide acetochlor and its degradation products were more toxic to bacteria than fungi.

Pesticides also influence the biochemical processes and enzymatic reactions in soil. Biochemical transformations of various elements (N, P, S, C) are carried out by soil microorganisms. These transformations are mineralization of organic matter, nitrogen fixation, nitrification, denitrification, ammonification etc. For example, soil when treated with prosulfuron, diminishes the processes of nitrification and denitrification. Application of fungicides like captan, apron and arrest have significantly decreased the rhizobia population in the seeds. This decreases the fixation of nitrogen in the soil.

Soil contains various enzymes, both extracellular and within microbial cells. These enzymes act as a biological index to understand the soil fertility and biological functions in soil. For example, glyphosate when applied in bunds can significantly diminish enzymatic activity. Similar effect is not seen when glyphosate is applied as granules. Many experiments have shown stimulating effects of enzymes on application of pesticides. For example, application of lower concentration of atrazine and northrin have shown to stimulate dehydrogenase activities of microbes.

EFFECT OF FERTILIZERS

Function of mineral fertilizer is to add nutrients to the soil to increase the soil organic carbon content. Response of fertilization is different for plant and microbial communities. It is seen that soil microorganisms are very sensitive to N, P and K fertilization. Nitrogen fertilizer have been a major contributor in increasing crop yield globally. But studies have shown that increasing nitrogen inputs generally decrease soil microbial biomass. Changes in microbial biomass depends on many factors such as pH, duration of application and soil management practices (Fig. 20.2).

Lu et al. (2010) for example, found that increasing N fertilization in unmanaged ecosystem have suppressed soil microorganism more than agricultural ecosystems. This is because higher productivity brought on by fertilizers, increases the carbon pool for microorganisms.

Long term studies by Geisseler and Scow (2014) showed that the addition of mineral fertilization significantly increased soil organic carbon content compared to the unfertilized control by an average of 12.8%. Population of microbial communities also increased by 15.1%. Since it is highly pH dependent, microbial diversity decreases in soil with pH below 5 and highest population in seen at pH 7.

Duration of application is very important to understand the changes in microbial diversity. Increase in microbial biomass is seen highest in studies that had been in place for at least 20 years.

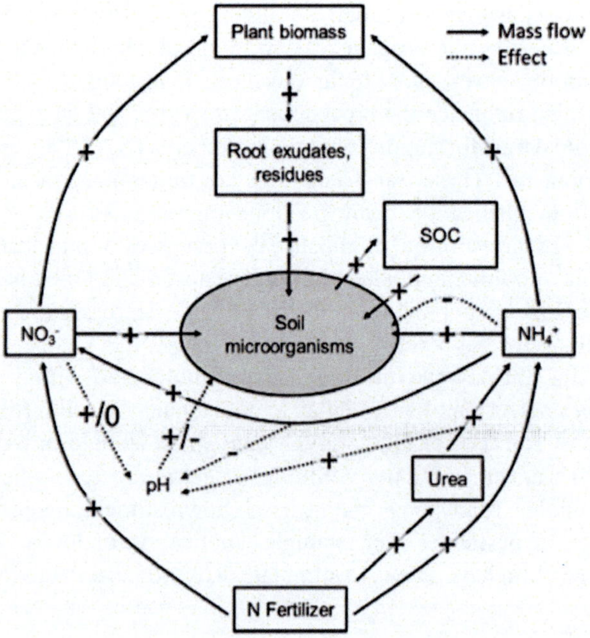

Fig. 20.2 Conceptual model of direct and indirect effects of mineral N fertilizer on microorganisms in agricultural soils (Geisseler and Scow, 2014).

Negative effect of fertilization to microbial communities has also been reported by many soil scientists. Studies by Klein et al. (1989) on western wheatgrass (*Agropyron smithii*) and wild rye (*Sitanion hystrix*) has shown that fertilization resulted in significant decrease in fungal hyphal lengths in the rhizospheres of both grasses. Wild rye also showed decrease in microbial biomass and soil organic carbon content in its rhizosphere.

Liang et al. (2020) studied the effect of biofertilizers and mineral fertilizers on rhizosphere bacterial communities of winter wheat. They found that biofertilizers increased bacterial diversity and the relative abundance of copiotrophic bacteria in rhizosphere soil, while nitrogen fertilizers reduced bacterial diversity but increased the relative abundance of oligotrophic bacteria.

CONCLUSION

Due to divergent findings of researches on relationship between soil microbiota and agrochemicals, it is very difficult to find their exact role. Some pesticides and fertilizers act as a energy source to the microorganisms but many have a degradative effect on them. Soil and environmental factors are responsible for the divergent relationship. Thus, understanding the mechanisms of underlying molecular responses in microbes in response to pesticides application could be helpful in assessing their risk and its consequent adverse impacts on soil biomass, enzymatic activities and biochemical reactions. Also, only 0.1% of total pesticide application reaches the target organisms. The rest accumulates as residues and pollute the environment. Biopesticide, which is a natural substance derived from microorganisms, have shown a significant result in controlling pests without harming the environment. These type of eco-friendly practices should be encouraged in future.

THINK FOR A WHILE

- How does different soil factors affect the application of pesticides on soil microorganisms?
- What are the processes that pesticides undergoes in soil?
- Why does application of pesticides tend to be more severe in coarse textured soil?
- How do pesticides influence the enzymatic reactions in soil? Give example.
- How does soil pH affect the microbiota on fertilizer application?
- Give an example of fungicide that has a negative effect on soil biota.

REFERENCES

Food and Agriculture Organization of the United Nations (2020). FAOSTAT Statistical Database. [Rome]: FAO.

Geisseler, D. and Scow, K.M. (2014). Long-term effects of mineral fertilizers on soil microorganisms – A review. *Soil Biology and Biochemistry*, Elsevier. **75:** 54–63.

Hussain, S., Siddique, T., Saleem, M., Arshad, M. and Khalid, A. (2009). Impact of Pesticides on Soil Microbial Diversity, Enzymes, and Biochemical Reactions, Advances in Agronomy. Elsevier Inc.

Klein, D.A., Frederick, B.A. and Redente, E.F. (1989). Fertilizer effects on soil microbial communities and organic matter in the rhizosphere of Sitanion hystrix and Agropyron smithii. *Arid Soil Res. Rehab.,* **3:** 397–404.

Liang, R., Hou, R., Li, J., Lyu, Y., Hang, S., Gong, H. and Ouyang, Z. (2020). Effects of Different Fertilizers on Rhizosphere Bacterial Communities of Winter Wheat in the North China Plain, Agronomy, MDPI., **10:** 1–12.

Lo, C.C. (2010). Effect of pesticides on soil microbial community. *Journal of Environmental Science and Health*. Taylor & Francis Group, LLC., **45:** 348–359.

Lu, X.K., Mo, J.M., Gilliam, F.S., Zhou, G.Y. and Fang, Y.T. (2010). Effects of experimental nitrogen additions on plant diversity in an old-growth tropical forest. *Glob. Change Biol.* **16:** 2688–2700.

Menon, P., Gopal, M. and Parsad, R. (2004). Influence of two insecticides, chlorpyrifos and quinalphos, on arginine ammonification and mineralizable nitrogen in two tropical soil types. *Journal of Agricultural and Food Chemistry,* **52(24):** 7370–7376.

Virag, J.A.I., Rolle, M.L., Reece, J., Hardouin, S., Feigl, E.O. and Murry, C.E. (2007). Fibroblast Growth Factor-2 Regulates Myocardial Infarct Repair: Effects on Cell Proliferation, Scar Contraction, and Ventricular Function. *The American Journal of Pathology*. **171:** 1431–1440.

Part V
Environment and Soil Quality

Chapter 21
Environmental Impacts

Traditionally soil has been considered as the medium for plant growth, source of different resources and foundation for civil structures. Studies of paleoclimate from paleosol have shown how earth's climate has changed for the past 50 million years. However, with increasing population, climate change, food security, soil degradation and decrease in soil biodiversity became major concerns (Fig. 21.1). Soil as mitigation of climate change through C sequestration, acting as bio-membrane for waste management and storehouse for interaction of different soil species has gained much attention in recent years. Sustainable agriculture includes maximizing production per unit area with much attention on improving soil quality and minimizing environmental pollution. In fact, excessive use of chemical fertilizers and conventional tillage has been responsible for decrease in soil quality causing soil degradation. Overuse of fertilizers, pesticides, free water, excessive tillage and overgrazing have resulted in current poor status of soil (Fig. 21.3).

CAUSES AND EFFECT OF SOIL DEGRADATION

Ecosystem factors of soil degradation includes climate (precipitation, temperature etc.), topography and vegetation. For example, soil erosion is severe when sloping lands and shallow soils are cultivated for row crop production in ecologically sensitive ecoregions like Himalayan ecosystem. Deforestation in ecologically sensitive regions (tropical rainforest) can cause decline in soil and water quality. These factors acts as catalysts which accentuates the soil degradation processes mentioned above. Anthropogenic activities such as agriculture, industry and urbanization alters the influence of these processes and factors. Also increase in demographic pressure, human needs, and other issues related to the human dimension further aggravates these processes.

S. Mukherjee, *Current Topics in Soil Science*,
https://doi.org/10.1007/978-3-030-92669-4_21

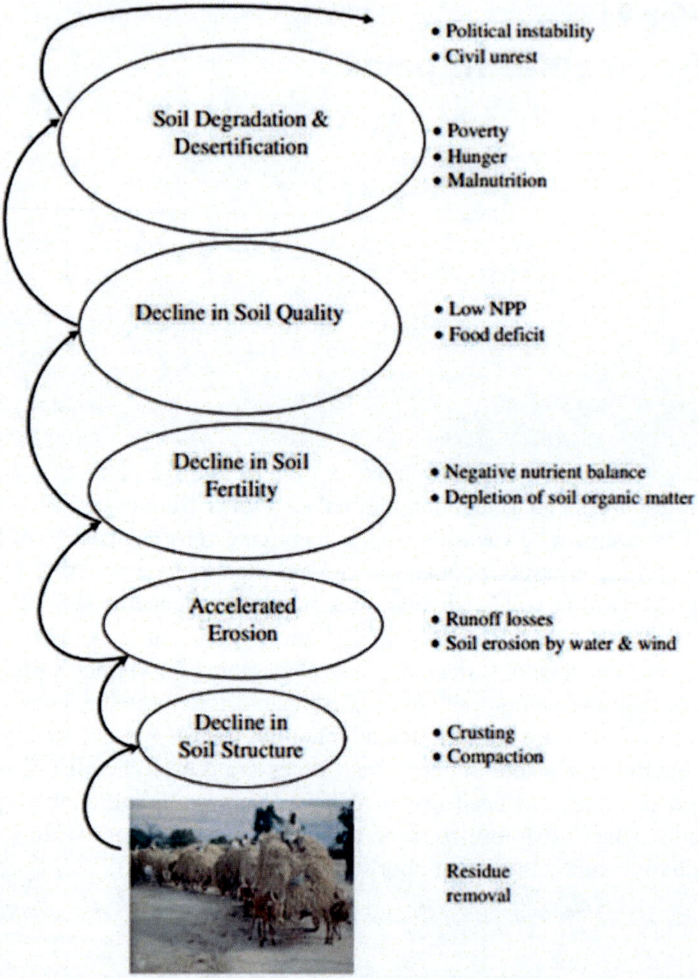

Fig. 21.1 The causes and effects of some major environmental concerns in soil.

FOOD SHORTAGE

The most direct effect of soil degradation is decline in productivity which is accentuated by loss in arable lands. The human population is doubling every 50 years and so is the increase in food supply. Excessive use of fertilizers and harsh chemicals, is causing permanent damage to the agricultural fields and the environment. It is seen that in non-irrigated dry lands of Africa, Asia and Latin America, excessive nutrient mining from soil causes crop yields to decline. With more demand, farmers are extending their low management practices into forest lands, vulnerable pasture lands or by shortening fallow periods. This mismanagement is not only causing nutrient imbalance but also habitat destruction and loss of biodiversity. Excessive

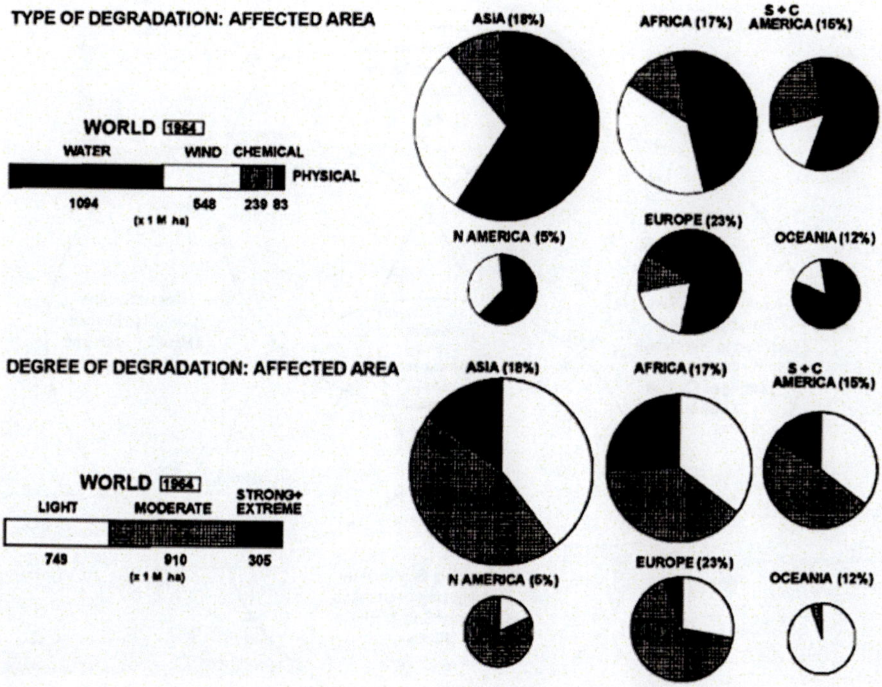

Fig. 21.2 Type and degree of degradation in different continents.

use of water for irrigation in arid and semiarid regions is causing salinization and water logging in soil. Acidification and pollution in soil is caused by inadequate nutrient supply and increase in fertlizers which causes contamination of waters and serious health hazard to both animals and humans.

It is seen that global per capita arable land area of 0.3 ha in 1985 declined to 0.23 ha in 2000, 0.15 ha by 2050, and 0.14 ha by 2100. Countries with per capita arable land area of less than 0.15 ha were 5 in 1960, 23 in 1990, and will be 50 in 2025 (Lal, 2006) (Fig. 21.2).

DEGRADATION AND CLIMATE CHANGE

Deforestation and agricultural practices are the main cause of greenhouse effect. Also excessive shift of climate pattern due to global warming is causing changes in precipitation and temperature patterns accelerating degradation in many areas or initiating degradation in new areas. Thus, degradation and climate change are intrinsically related to each other.

Soil which act as the largest carbon reservoir in terrestrial ecosystem, holds carbon in the form of organic matter. This organic carbon is determined by soil temperature and moisture regime, internal drainage, textural composition, and clay

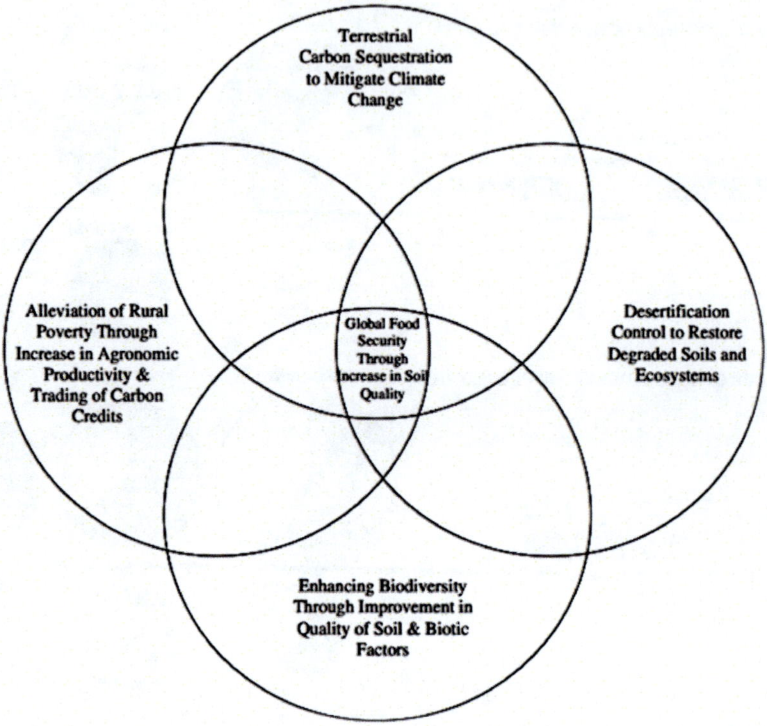

Fig. 21.3 Interaction between desertification control, biodiversity improvement, climate change mitigation, and food security which is improved through improvement in soil quality, increase in availability of water resources, strengthening of elemental cycling, and enhancement of bioturbation in the rhizosphere.

mineralogy. Fertile soils contains about 4–6% of carbon in weight. Thus, degradation in the form of deforestation, agricultural practices, soil erosion, salinization and fertility decline is causing carbon from soil to be released in atmosphere resulting in greenhouse effect. It is seen that from 1860 to 1980, the net decline of the carbon pool in plant biomass is 22 Gt, and in the soil 5 Gt. In fact due to degradation many soils throughout the world do not achieve their potential carbon-carrying capacity. This is particularly the case in most arid and semiarid regions of the developing world, where overgrazing, fuel-wood collection, and opportunity cropping have seriously depleted the limited stock of organic matter (Pauw and Zoebisch, 2006).

In some cases degradation do not lead to carbon loss. There is evidence that in geological past, an increased atmospheric CO_2, coupled with warmer temperature, higher rainfall, and increased biological activity could lead to higher concentration of reactive bicarbonate (HCO_3) in soil solution, producing more deeply weathered soil profiles rather than atmospheric carbon.

SOIL MANAGEMENT PRACTICES

Much work is done to keep carbon in soil through restoration of peatlands and wetlands, afforestation etc. Organic fertilizers such as manures, compost, biosolids, green manure and cover crops are used instead of chemical fertilizers to reduce environmental pollution. Supplying nutrients in proper amount and in right time is important for soil health which can be done by nutrient management. Educating farmers on fertility and nutrient management is the first step to ensure soil longevity for future generations. With advancement of technologies, scientist can use GPS to track fertilizer and water uses in a precise area. This will help prevent wastage of resources and labour.

TILLAGE AND CROP ROTATION

No Tillage

This tillage is a type of conservational tillage that maintains at least 30% of the soil cover (1000 kg ha^{-1}) of crop residues after planting the next crop. It helps in controlling soil erosion and conserve water. This method includes sweeping, chiseling, paraploughing, subsoiling, slit and strip rotary.

With time, tillage disturbs soil surfaces by reducing SOC concentration with increase in bulk density. Tillage systems also affect soil biological activity, soil water and temperature regimes, and the accessibility of C and N substrates that influence elemental cycling. No tillage significantly help increasing soil organic carbon, microbial biomass, and potentially mineralizable N in soil compared to conventional tillage. It helps in reducing the soil temperature which results in better crop performance in warm region. Greatest advantage in no tillage is lower equipment inventories, low equipment maintenance, and lower energy and labour requirements. No tillage is one of the most important factor in crop management for SOC sequestration. It is expected that if one third of the cultivated land practice no tillage, SOC sequestration will improve drastically.

Crop Rotation

This refers to growing of different crops sequentially in the same area across their growing season. Monocultures or growing the same crop in the same area for many years can result in depletion of many nutrients and increase in competitive pest and weed community. Thus, crop rotation in combination with no tillage can improve soil quality by reducing soil erosion, retention of C and N in soil, enhance SOM, and sequester SOC. The changing of crops in a sequence decreases the population level of pests by interrupting pest life cycles and pest habitat. Studies conducted over years has concluded that crop yield is significantly increased with crop rotation compared to monocultures. For example, long-term experiment by Buyanovsky and Wagner (1998) in Missouri reported that monoculture of wheat with N fertilization accumulated 50 gCm^2 y^{-1} compared with 150 gCm^2 y^{-1} by corn-wheat-clover rotation with manuring and N fertilization. But crop rotation requires a great deal of planning. Choosing of right crop depends on soil type, topography, climate and irrigation in addition to conditions that may change dramatically from year to year such as weather, market and labour supply.

COVER CROPS AND GREEN MANURE

Cover crop refers to crops that protects and improve soils between normal crop production. It is grown primarily for the benefit of soil rather than crop yield. Cover crops are commonly grasses and legumes which helps in controlling soil erosion, suppress weed production, pest control, and improve soil fertility and groundwater quality. Cover crop is grown between cash crops to prepare the soil for the next important yield.

In contrast, green manures are crops that are grown and incorporated into the soil to improve soil fertility (Fig. 21.4). A cover crop may be referred as a green manure when ploughed under and incorporated into the soil. Cover crops increases soil quality while green manures works on soil productivity. Green manures provide additional nutrients and organic matter to soil without using chemical fertilizers that

Fig. 21.4 Green manure (Dhaincha) for cultivation of cotton.

harm the environment. It also helps in improving soil structure by increasing their water retention and drainage capabilities. For example, leguminous green manures such as clover and vetch contain nitrogen-fixing symbiotic bacteria in root nodules that fix atmospheric nitrogen in a form that plants can use. With appropriate choice of species and proper management, cover crops can increase SOC sequestration. Growing cover crops may contribute 25 to 50 kg N ha^{-1} by small grains and 36 to 226 kg N ha^{-1} by legumes (Reeves, 1994).

PRECISION FARMING

Precision farming is technology based approach to farm management with the help of GPS (Global Positioning System) that observes, measures and analyzes the need of individual fields and crops. Soil's spatial variability, computerized soil maps, aerial photographs, spot satellite images of the field, and grid maps detailing yield potentials, soil limitations, and pest histories are widely used to identify areas of similar soil fertility (Pierce and Lal, 1991). Precision farming reduces the cost and wastage of resources because it tells the farmer exactly where nutrients or pests are required in a field. It also helps in water conservation and favours SOC sequestration (Fig. 21.5). With the help of real time sensors, variables such as crop yield, topography, organic matter content, moisture levels, nitrogen levels, pH, EC, Mg, K etc. can be measured. With the advancement of technology, unmanned vehicles, drones are used to capture imagery that can be enhanced to be used for correlating crop health to topography.

MANURE, COMPOST AND BIOSOLIDS

Manure, compost and biosolids are organic residuals that are used to add organic matter and nutrients into the soil. Manures are animal waste that is generated in animal farms or dairy barns. They are rich in macro- and micronutrients and need little management as they decompose rapidly. The amount of N, P and K in manures replace upto 45% of chemical fertilizers. Composting is the process of aerobic decomposition of organic matter at temperature range of 40–65°C. Compost are rich in humus and also act as a pesticide, control erosion, land and stream reclamation, wetlandconstruction, and as landfill cover. Ideal factors for compost are C : N ratio of 25 : 1–30 : 1, 50–60% moisture content, oxygen concentration >5%, pH 6.5–8.0, and particle size of 3–13 mm. Organic byproducts from waste water treatment results in biosolids. All these increase the C concentration in soil resulting in increase in SOC. It is seen that even if C in soil is not increased significantly, these organic amendments increases microbial biomass, rhizodeposition, and plant growth. Management of manure, compost and biosolids, if not done properly, can lead to environmental pollution. In some parts of India, farmers use cow dung as

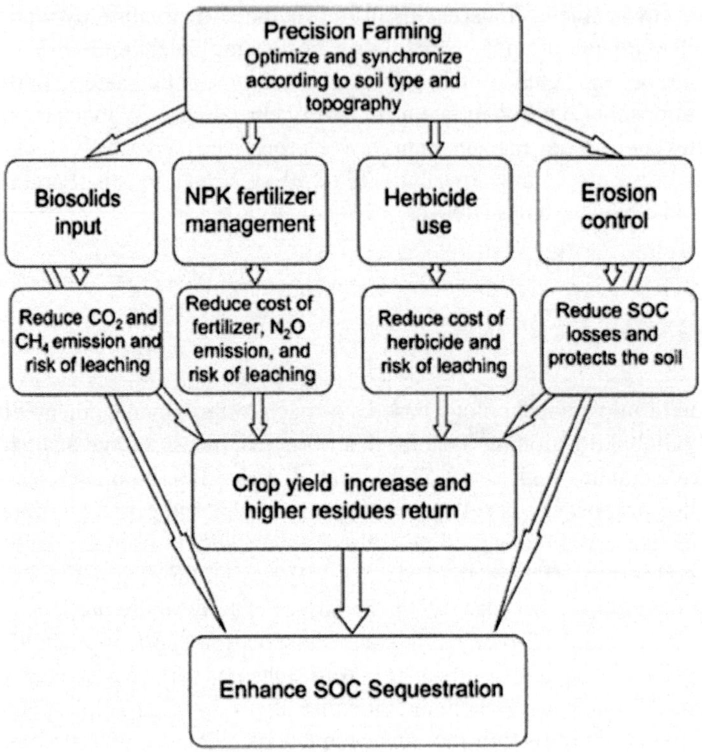

Fig. 21.5 Precision farming and its effect.

fuel. Thus, it is a serious challenge to encourage them for using these organic amendments as fertilizers.

WASTE MANAGEMENT

With increasing population and extraction of resources, there is large generation of waste which has concerned scientists, governmental and environmental agencies for their disposal destination. Thus, recycling of waste has become very important to ensure water, nutrients and food for future. Use of biosolids as organic amendments in place of chemical fertilizers and conversion to methane and ethanol from them have helped a significant portion of wastes to be recycled.

Appropriate soil having good drainage, absence of impermeable layer in the vadose zone, high activity and species diversity of macro- and micro-organisms in the surface layer, highly aggregated and stable structure can act as a biomembrane for filtering biodegradable industrial wastes. Underlying geological strata is also used for storing nuclear waste. Management of waste based on degree of hazard and determining the soil assimilative capacity for specific wastes and chemicals and how that assimilative capacity can be managed in a positive manner, has become prime study for soil scientists. This can be done by efficient use of resources, reducing the amount of waste generated by households, industry, resource recovery including reuse, recycling, reprocessing, and energy recovery, consistent with the most efficient use of the recovered resources and disposal including management of all disposal options in the most environmentally responsible manner.

CONCLUSION

Access to adequate and balanced food and clean drinking water are the most basic human rights. To ensure this, it is our duty to be aware and educate others. These issues should be addressed by scientists, activists, environmentalists and reach out to other disciplines while strengthening and advancing the science of soil and its dynamics in an ever-changing physical, social, economical and political climate. Policies need to be adaptive and make allowance for the reality of complex evolving systems, avoiding the pitfalls of a command and response strategy.

THINK FOR A WHILE

- What are the current major concerns among soil scientists?
- Is there any way where soil degradation has positive effect on climate change?
- How does degradation cause salinization and acidification in soil?
- What are the advantages of no tillage?
- Why crop rotation is better than monoculture?
- What is the difference between cover crops and green manure?
- What are the advantages of precision farming?

REFERENCES

Buyanovsky, G.A. and Wagner, G. (1998). Changing role of cultivated land in the global carbon cycle. *Biology and Fertility of Soils,* **27:** 242–245.

Lal, R. (2006). Degradation and Quality. Taylor & Francis.

Pauw, E.D. and Zoebisch, M.A. (2006). Degradation and the Greenhouse Effect. Taylor & Francis.

Pierce, F.J. and Lal, R. 1991. Soil management in the 21st century. *In:* Soil Management for Sustainability, pp. 175–179. Lal, R. and Pierce, F.J., Eds., Papers from the workshop in Edmonton, Alberta, Aug. 1989. Soil and Water Conservation Society, Ankeny, Iowa, USA.

Reeves, D.W. (1994). Cover Crops and Rotations. *In:* Hatfield, J.L. and Stewart, B.A., Eds., Crops Residue Management. Advances in Soil Science, Lewis Publishers, Boca Raton, 125–172.

Chapter 22
Effect of Climate Change on Soil Ecosystem

Soil biodiversity includes soil microorganisms and soil fauna, their functions and interactions between them. Controlling factors of soil ecology are: acting as the primary driving agents of nutrient cycles, regulating the dynamics of soil organic matter, modifying soil physical properties, enhancing the amount and efficiency of nutrient acquisition by the vegetation, enhancing plant health, greenhouse gas emission and carbon sequestration. These interactions are also reflected in the food web.

Nature and human activities strongly influence the ecosystem. Natural activities include both long term (erosion etc.) and short term (natural disasters like earthquake, landslide etc.). In human activities, long term include transformation of land for agriculture and short term includes variety of waste disposal (i.e industrial disposal and municipal garbage) which affects both soil and water system of the landscape. This is degrading the biodiversity worldwide. Atmospheric CO_2 variation affects the climate change positively if it is reduced and negatively when enhanced. Thus, present human activity management tries to reduce the CO_2 emission. Also with elevated temperature, N mineralization also increases, enhancing the nutrition in plants. This can disturb the ecosystem at all levels. Present status of climate change management will be discussed.

CLIMATE CHANGE

Earth's atmosphere which contains gases like oxygen, ozone and nitrogen, absorbs harmful ultraviolet radiation from the sun. Earth's surface absorbs the incoming solar radiation and re-emits back to space as infrared radiation. Greenhouse gases like water vapour, methane, nitrous oxide, ozone and chlorofluorocarbons (CFC) trap the longer wavelength radiation emitted from the earth's surface thereby maintaining the mean surface temperature at 15°C. This phenomenon is called the *greenhouse effect*. With increasing greenhouse gases in the atmosphere, the temperature of earth is also rising causing *global warming*.

S. Mukherjee, *Current Topics in Soil Science*,
https://doi.org/10.1007/978-3-030-92669-4_22

Another rising concern is that gases such as CFC's (man-made compounds), are diffusing from the troposphere into the stratosphere, destroying the ozone layer responsible for protecting us from the harmful UV rays.

The CO_2, one of the most important greenhouse gas and the main cause of global warming, is increasing in concentration due to burning of fossil fuels. Oxidation of soil organic matter also contribute significant amount of CO_2 into the atmosphere.

Another greenhouse gas, methane is produced by bacteria during anaerobic respiration which only occurs under strongly reducing conditions such as natural wetlands. Rice paddies also contribute significant concentration of methane as it is grown in submerged soils. In fact, due to increasing demand of rice, large areas of lands are cultivated for rice increasing the concentration of methane at the rate of about 0.9% per year. This is more rapid than increasing concentration of CO_2.

Nitrous oxide (N_2O) is produced from soils during the process nutrient cycling of nitrogen. Nitrogen fertilizers too add a significant amount. It is estimated that the production of N_2O from N fertilizer is about 1.5% of that applied. Annual use of N fertilizers is more than 60 million tons.

It is clear that soils play a major role in climate change. The lifetime of these greenhouse gases in the atmosphere ranges from 10 years for CH_4 to 150 years for N_2O. Increasing concentration of these gases is increasing the global temperature at an average rate of 0.17°C per decade since 1970. Climate change has a profound effect on soil organisms, quality and quantity of organic matter including their density, biomass, diversity, activity, rates of consumption, life history parameters and migration ability which is discussed further in this chapter.

EFFECT OF ELEVATED CO_2 ON SOIL BIOTA

The direct effect of elevated CO_2 on soil biota is negligible. This is because these organisms are adapted to the already high CO_2 concentration in the soil. Indirect effects such as changes in litter quality and quantity, and increase in global temperature can affect their density, biomass and the composition of the communities (Fig. 22.1).

Several studies have shown that rise in concentration of CO_2 has increased plant productivity by increase in photosynthesis. Due to this, plant derived C concentration returned to soil has also increased, stimulating soil microbial growth and activity. However, this is only seen in nutrient rich condition where N is not limited. Possibility of microbial growth in nutrient rich condition due to elevated CO_2 concentration has high mixed opinions. This is due to the fact that some components of the microbial community are regulated by predation versus resource supply. Any potential increase in microbial biomass, or the abundance of a particular microbial group, could be dampened by increased consumption by microbial-feeding soil animals (Wardle, 2002).

However increasing growth of soil fauna on elevated CO_2 concentration has mixed responses. For example, Freckman et al. (1991) found that the density of

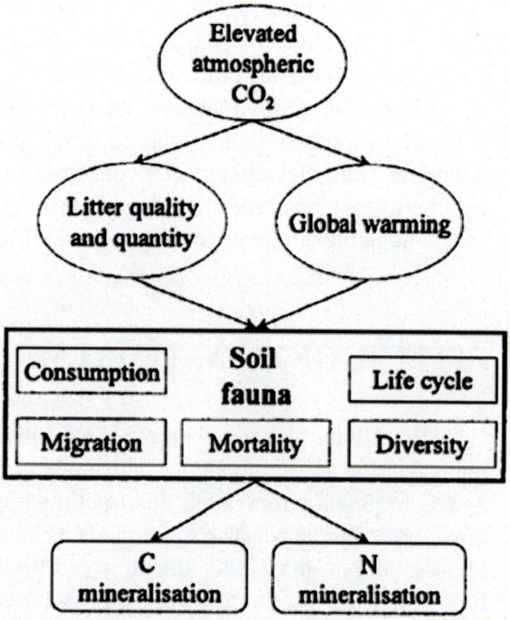

Fig. 22.1 Conceptual flowchart of the effect of elevated atmospheric CO_2 on soil fauna.

nematodes, in soils of western wheatgrass and blue grama grasses, were unaffected after 2 years exposure to elevated concentration of atmospheric CO_2 whereas population densities of enchytraeid worms were found to increase in response to elevated CO_2 in grassland (Newton et al., 1995; Yeates et al., 1997). Arnone and Zaller (1997) observed that earthworm surface cast production was 35% greater over one year in open top and open bottom rings under elevated CO_2 concentrations. This in-turn increases the structure of soil like increase in pore volume, water-holding capacity, more water-stable aggregates and greater aeration. Conditions where changes in microbial density and community structure occurs in response to elevated CO_2 concentration, have the potential to influence higher trophic levels.

EFFECT OF ELEVATED CO_2 ON NUTRIENT AVAILABILITY

Increase in CO_2 concentration can influence microbial growth and community structure. This influences the rate of decomposition, nutrient mineralization and nutrient availability to plants. Also under nutrient rich condition elevated CO_2 concentration can increase photosynthesis in plants. Rate of increase in photosynthesis causes elevation in soil C and N dynamics promoting soil N availability. But in some case negative impact too can take place. For example, Díaz et al. (1993) showed that while elevated CO_2 concentration increased the amount of C returned to soil via roots, an enlarged microbial biomass led to net immobilization of N, thereby limiting the availability of N to plants. To determine whether the impact is positive or negative, depends on C : N ratio of the substrate. In general, CO_2 elevation increases the C : N ratio, increasing N immobilization by microbes and limiting N to plants. Limited N increases the competition between plants and microbes for N thereby suppressing microbial decomposition and increasing ecosystem C accumulation. Thus, elevation in CO_2 concentration has influenced partitioning of nutrients between plants and microbes disturbing the base level ecosystem. This will have a huge impact on larger tropic levels (Fig. 22.2).

EFFECTS OF ELEVATED TEMPERATURE

Growth, activity and composition of microbial communities are susceptible to two main factors, viz. temperature and moisture. Although there is a net increase of 1–3°C in global temperature due to climate change, it does not have a significant effect on soil biota because soil organisms have a wide range of temperature optima. Despite of that, many soil functions can get affected especially for the organisms living in extreme climate. Peatland and arctic ecosystems are said to be most affected by rising temperature.

Peatlands contains one-third of global carbon stock. With elevated global temperature, decomposition of peatlands is causing carbon to be released in gaseous and aqueous form. This is supported by the rising Dissolved Organic Carbon (DOC) concentrations in aquatic ecosystems that receive drainage water from peatlands in

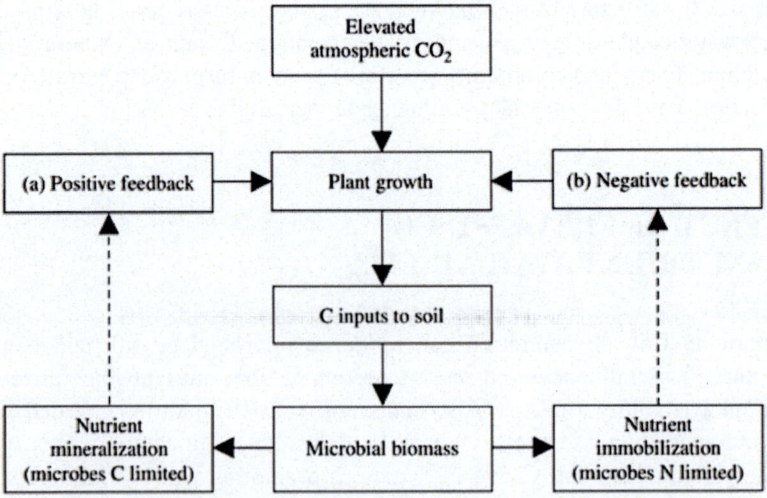

Fig. 22.2 The potential effects of elevated atmospheric CO_2 on soil C inputs and microbial biomass, leading to either: (a) positive or (b) negative feedback on nutrient availability and plant growth (Zak et al., 1993 and Díaz et al., 1993).

the United Kingdom. Enhanced decomposition is caused by increase in microbial activity due to rising temperature.

With warming in arctic ecosystem, increase in organic matter decomposition will result in carbon loss. N mineralization are very low in arctic condition. Increase in temperature can lead to increase in rate of N mineralization leading to increased nutrient supply to plants. This in-turn is predicted to change plant community composition in favour of species with greater nutrient uptake rates, such as graminoids and deciduous shrubs, at the expense of plants with lower uptake rates and growth rates, such as mosses and evergreen shrubs (Chapin et al., 1996; Press et al., 1998). Since microbes hold 7% of the total N pool and 35% of the total P pool in arctic ecosystem, it is possible that the competition among plants and microbes will also increase. For example, warming increased growth of *Dryas octopetala*, a dominant arctic plant, but reduced its foliar N content (and increased its C : N ratio), which would presumably reduce decomposition, lowering nutrient availability, and possibly lowering plant N levels and growth in subsequent years (Welker et al., 1997). Thus, changing vegetation with climate change will also affect litter quality and quantity, subsequently changing the soil biological properties. For example, Hobbie (1996) pointed that deciduous shrubs affected by rising temperature, produce large amounts of low quality, woody litter that decomposes slowly, potentially leading to C accumulation on the soil surface affecting the hydrology of soil. Snow acts as an insulator of soil. Elevated temperature is decreasing the snow line resulting in colder soil temperature, more extensive freezing, and an increase in the frequency of freeze-thaw cycles over the winter months.

CONCLUSION

Soil ecosystem is directly and indirectly affected by climate change. The most common effects are increase in photosynthesis in plants, microbial growth, altering soil community structure and nutrient availability etc. Effect in carbon and nitrogen concentration in soil is most important which further alters the dynamics of soil ecosystem. Also many of these plant-soil interactions are species-specific and are greatly affected by temperature and moisture. Climate change also alters the flora-fauna dynamics especially increasing competition between them for nutrients that may also affect the terrestrial food web.

THINK FOR A WHILE

- Why the effect of increased CO_2 concentration on soil biota is negligible?
- Why during elevated CO_2 concentration, microbial growth in nutrient rich condition is not always true?
- What is the effect of soil N availability on elevated CO_2 concentration?

- What happens if C : N increases in soil?
- How does decomposition rate increase with increasing temperature?
- How does litter quality get affected by global warming? What will be its effect on C and N content in soil?

REFERENCES

Arnone, J and Zaller, J. (1997). Activity of surface-casting earthworms in a calcerous grassland under elevated atmospheric CO_2. *Oecologia*, **111:** 249–254.

Chapin III, F.S., Torn, M.S. and Tateno, M. (1996). Principles of Ecosystem Sustainability. *The American Naturalist.* **148:** 1016–1037.

Díaz, S., Harris, J.A. and Mcpherson, E. (1993). Evidence of a feedback mechanism limiting plant response to elevated carbon dioxide. *Nature.* **364:** 616–617.

Freckman, D.W., Barker, K. R., Coleman, D.C., Acra, M., Dyer, M.I., Strain, B.R. aand McNaughton, S. J. (1991). The Use of the 11C Technique to Measure Plant Responses to Herbivorous Soil Nematodes. *Functional Ecology,* **5**(6): 810–818.

Hobbie, S.E. (1996). Temperature and plant species control over litter decomposition in Alaskan tundra. *Ecol. Monogr.* **66:** 503–522.

Newton, P.C.D., Clark, H., Bell, C.C., Glasgow, E.M., Tate, K.R., Ross D.J., Yeates, G.W. and Saggar, S. (1995). Plant growth and soil processes in temperate grassland communities at elevated CO_2. *Journal of Biogeography,* **22:** 235–240.

Press, M. C., Potter, J. A., Burke, M.J.W., Callaghan, T.V. and Lee, J.A. (1998). Responses of a sub-Arctic dwarf shrub heath community to simulated environmental change. *J. Ecol.* **86:** 315–327.

Wardle, D.A. (2002). Communities and ecosystems: Linking the aboveground and belowground components. 1st edn. Princeton University Press, Princeton.

Welker, J. M., Molau, U., Parsons, A.N. et al. (1997). Response of Dryas octopetala to ITEX manipulations: A synthesis with circumpolar comparisons. *Global Change Biology 3* **(suppl. 1):** 61–73.

Yeates, K.O., Taylor, H.G., Drotar, D., Wade, S., Stancin, T. and Klein, S. (1997). Preinjury family environment as a determinant of recovery from traumatic brain injury in school-age children. *Journal of the International Neuropsychological Society,* **3:** 617–630.

Zak, D.R., Pregitzer, K.S., Curtis, P.S., Teeri, J.A., Fogel, R. and Randlett. D.L. (1993). Elevated atmospheric CO_2 and feedback between C and N cycles. *Plant and Soil.* **151:** 105–117.

Chapter 23
Soil Carbon Sequestration

Carbon sequestration is the process by which carbon dioxide is removed, captured or stored from the atmosphere to slow down pollution and to mitigate or reverse global warming. The most important service provided by the ecosystem is soil carbon storage which results from interactions of many ecological processes (photosynthesis, respiration and decomposition). Soil organic matter (SOM) is the key component responsible for physical, chemical and biological properties of soil. This includes soil quality, structure, chemical composition, productivity etc. With increasing population and industrial revolution, burning of fossil fuels have led to increase of carbon in atmosphere. Loss of carbon is taking place due to agriculture and improper land management. This chapter mainly deals with the processes of mitigation of global warming with the help of carbon sequestration.

SOIL CARBON AND CARBON CYCLE

In terrestrial ecosystem 80% of total carbon is found in soil. This amounts to nearly 2500 GT. Carbon in soil can be inorganic which comes from weathering of carbonate minerals (gypsum, calcite). Levels of soil organic carbon (SOC) depends on the amount of SOM. It also depends on climatic factors like temperature and precipitation. For example, arctic climates have high SOC level as rates of photosynthesis exceed decomposition while arid climates have low SOC. During summer when temperature and moisture levels are highest, temperate ecosystems can have high primary productivity. With low temperatures during the rest of the year, decomposition rates become slow such that organic matter slowly builds up over time.

The CO_2 is fixed by plants by photosynthesis. Amount of organic carbon input depends on organisms living in plant roots (symbiotic relationship) and also amount of litter from dead plants. Loss of C takes place by decomposition of biomass by microbes through the process of microbial respiration (Fig. 23.1). Not all carbon is lost, some is stored as humus. Carbon loss can also take place due to soil erosion

S. Mukherjee, *Current Topics in Soil Science*,
https://doi.org/10.1007/978-3-030-92669-4_23

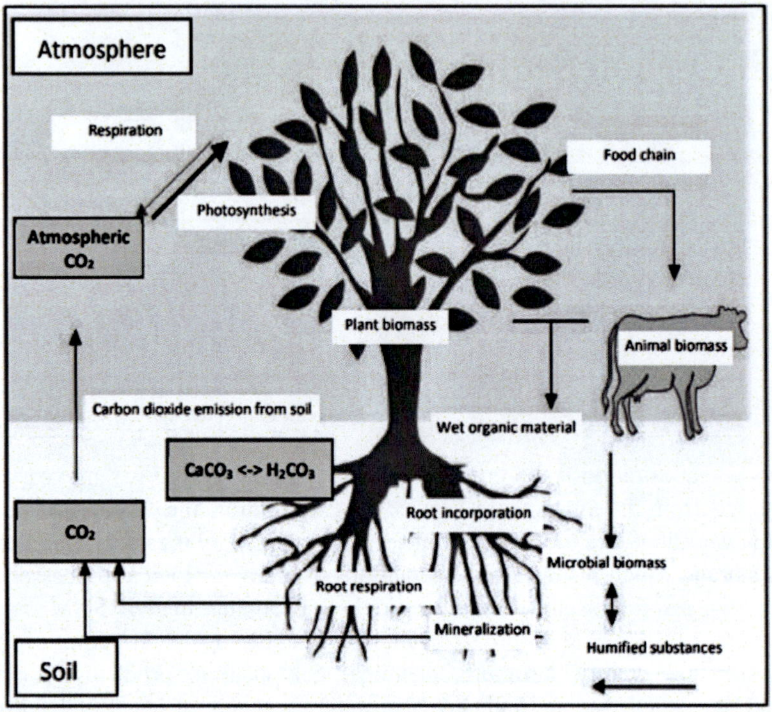

Fig. 23.1 The carbon cycle and interaction of plant, soil and atmosphere.

and leaching of carbon into groundwater. Carbon cycle is healthy when both input and output of carbon is balanced.

MITIGATION

Carbon sequestration can take place by biological, physical or chemical processes. It can be natural or artificial (man-made). The main goal of carbon sequestration is removing carbon from the atmosphere and storing carbon in soil (Fig. 23.2).

- **Removing carbon from air:** Biosphere has the ability to increase carbon assimilation (photosynthesis) when carbon concentration in the atmosphere increases. This helps in reducing or slowing down the process of greenhouse effect. Plants also have the ability to enhance water use efficiency increasing the total global plant population. This ability helps plants to live and successfully reproduce in areas where it was formerly too dry for them to survive.

Hence, to speed up the sequestration process, artificial or man-made techniques can be used like afforestation, reforestation or forest management, i.e. increasing the plant population by planting more trees (Table 23.1). Even there are some

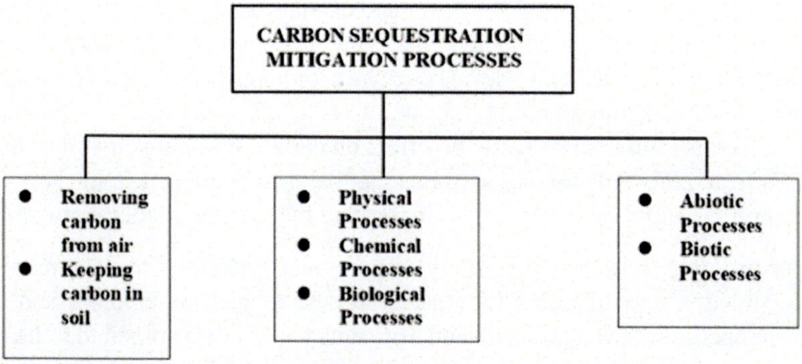

Fig. 23.2 Processes for mitigating carbon from atmosphere and storing carbon in soil.

Table 23.1 Potential rates of carbon sequestration (kg C/ha/yr) due to different land management practices

Land management practices	*Rates of carbon sequestration (kg C/ha/yr)*
Improved rangeland management	50–150
Improved pastureland management	100–200
Commercial fertilizer applications	200–500
Manure applications	100–300
Improved grazing management	300–1300
Nitrogen fertilization of mountain meadows	100–200
Restoration of eroded soils	50–200
Restoration of mined lands	1000–3000
Conversion of cropland to pasture	400–1200
Conversion of cropland to natural vegetation	600–900
Conversion from conventional tillage to conservational tillage	
No till	500
Mulch till	500
Ridge till	500

Adapted from Lal et al. (1998).

species of trees in tropical rainforests that can live more than thousand years. Protecting such species can help remove CO_2 for a long time. Agriculture and proper land use management practices can also help mitigate CO_2 from atmosphere. It can be done by increasing acreage of land devoted to growing crops. But simultaneously it can also increase carbon loss from soil by excessive tillage practice. Transition to less intensive tillage (conservational tillage) in place of long history of conventional management can help rebuild previously lost stores of soil organic matter.

- **Keeping carbon in soil:** The main target of carbon sequestration is to keep at least the same percentage of carbon in the soil as has historically been the case and especially in the face of potential global warming. Carbon is added to soil by

CO_2-enhanced root growth, turnover and exudation, as well as from CO_2-induced increase in leaf litter and other decaying plant parts. Although studies have shown that increase in CO_2 levels does little or no effect in decomposition rate of plant litter. Like temperature, elevated CO_2 levels too decreases decomposition rate. Temperature have mostly no effect on carbon loss but sometimes decreases CO_2 from soil. This mainly suggests that balance is created in the ecosystem to control the increase in C level by increasing the organic matter content in soil.

For example, in a two-year study of perennial ryegrass grown at ambient and twice-ambient atmospheric CO_2 concentrations, as well as ambient and ambient +3°C temperature levels, Casella and Soussana (1997) determined that the elevated CO_2 increased soil carbon storage by 32% and 96% at low and high levels of soil nitrogen supply, respectively, with no significant increase in temperature effect (Idso and Idso, 2006). This study suggests that to retain the excess carbon due to increase in CO_2 levels, there is a increase in plant population.

PHYSICAL PROCESSES

- **Bioenergy Production:** Bioenergy is the process by which energy is extracted from biomass in the form of liquid, solid or gase. It can be used for transport, heating, electricity production, combined heat and power generation, cooking through the processes of combustion, fermentation, pyrolysis etc. The carbon in the biomass is extracted from CO_2 in the atmosphere thereby reducing greenhouse gas emission. Some of the carbon in the biomass is converted to CO_2 or *biochar* which can then be stored by geologic sequestration or land application. Biochar is charcoal created by the process of pyrolysis from biomass waste. This charcoal is mainly used in landfill to increase the carbon stock in soil for long term mitigation of climate change. Burial of other biomasses such as trees also mimics the production of fossil fuels in long term.
- **Geological Sequestration:** This is the process by which CO_2 is generated from the emission of fossil fuels and industries, is captured and then stored away from the atmosphere for very long time. This method is also called Carbon Capture and Storage (CCS). The concept is that once CO_2 is captured from a point source (e.g cement industry), it would be compressed to $\approx$100 bar to form a supercritical fluid. In this fluid form, the CO_2 is transported via pipeline to the place of underground storage (at least 1 km) such as depleted oil and gas reservoirs, saline formations, or deep, un-minable coal beds, where it would be stable for hundreds to millions of years. At these storage conditions, the density of supercritical CO_2 is 600 to 800 kg/m^3. Determining the site for storage is very important. Ideally, rocks with high porosity and permeability and absence of faults are good storage sites for CO_2. Sandstone or limestone with shale as caprock with absence of faults is also an ideal storage site.

- **Ocean Storage:** Ocean is the largest carbon sink on earth storing 39000 Gt of C and it has the potential to store thousands of gigatonnes more. CO_2 is currently emitted at the rate of 10 Gt/year where ocean absorbs 2.4 Gt/year. Thus, scientist were encouraged to inject CO_2 directly into the ocean. But it comes with great risk to the marine biota. Since the site where CO_2 will be injected can become acidic by reduction of pH by CO_2 which can be toxic to the marine lives. Elevated atmospheric CO_2 impacts phytoplankton and coral reef organisms. The rate of calcification of corals macroalgae and phytoplankton decreases with increasing CO_2 concentration in ocean (Langdon et al., 2000). It is believed that liquefied CO_2 injected at 3000 m depth could be enough for stable injection of CO_2 in ocean but the environmental impact remains the most essential and controversial problem of ocean CO_2 sequestration.

CHEMICAL PROCESSES

- **Mineral Carbonation:** This is an inorganic form of sequestration by which carbon, in the form of CO_2, is removed from the atmosphere by chemical processes, and stored in stable carbonate mineral forms.Weathering causes exothermic reaction of carbon dioxide with abundantly available metal oxides such as magnesium oxide (MgO) or calcium oxide (CaO) to form stable carbonates. These minerals are abundantly available in silicate rocks such as serpentine, forsterite etc. Formation of limestone caves is an example of CO_2 extraction from atmosphere. Reaction rate increases with increase in temperature and pressure.

$$CaO + CO_2 \rightarrow CaCO_3$$

$$MgO + CO_2 \rightarrow MgCO_3$$

- **Ocean Sequestration:** Natural ocean absorbs a lot of CO_2 from atmosphere. This happens when CO_2 mixes with seawater and injects into hot basaltic rocks underground reacting to form solid carbonate minerals releasing Ca^{+2} and Mg^{+2}. Thus, underground basalt act as a large carbon sink in ocean and risk of leakage is minimum due to formation of hydrates.

BIOLOGICAL PROCESSES

- **Restoring Biological Carbon Sinks:** Although biological processes have already been discussed above, the most important man-made process is to restore and enhance the largest biological carbon sinks such as peatlands, wetlands forestry etc.

Although wetlands comprise only 6% of world's land, it contains 14.5% of world's soil carbon. Peatlands, which is also a type of wetland, sequesters 0.37 GT of CO_2 every year. They sequester carbon from the atmosphere through plant photosynthesis and by acting as sediment traps for runoff. This carbon is held in the living vegetation, litter, peats, organic soils as well as sediments that have built up over thousands of years. Thus, restoring these carbon sinks become vital for mitigating climate change and can be done by:

1. Reducing wetland drainage that leads to dewatering of wetlands.
2. Control peat harvesting that removes carbon from wetlands.
3. Restore revegetation to prevent proliferation of invasive species due to climate change.
4. Prevent fire risk which mainly happens due to dewatering of wetlands.

    - **Iron Fertilization:** Phytoplankton photosynthesis is one of the mechanisms which fixes approximately 45 Pg C yr^{-1} (Falkowski et al., 2000). One of the limiting factor of phytoplankton growth is Fe. This oceanic iron fertilization is a geoengineering technique that involves intentional induction of iron-rich deposits into oceans to enhance biological productivity. Although natural fertilization of nutrients takes place in ocean, such as upwelling of currents, but artificial iron fertilization can have serious effect on marine ecosystem. It can significantly disrupt ocean's nutrient balance and cause major complications in the food cycle for other marine organisms.

CONCLUSION

With human interference, the atmospheric CO_2 concentration continue to rise. Both natural and man-made processes are involved to decrease this trend. Natural processes involve increase in plant population by enhancing water and nutrient uptake. Man-made processes include afforestation, deforestation, forestry etc. Proper land use and agricultural management can also help decrease carbon from air. There are many geoengineering techniques used for carbon sequestration involving use of bioenergy, storing liquified CO_2 underground, ocean storage, iron fertilization etc., but it comes with many environmental risks. Hence, use of these techniques are of a great scientific debate.

THINK FOR A WHILE

- What are the ecological processes responsible for carbon storage in soil?
- What is the importance of SOM in carbon sequestration?
- Why arctic climate have high carbon content in soil?
- How does nature react to increasing CO_2 concentration in air?

- What is biochar? How it is used to mitigate climate change?
- What is an ideal site for carbon storage underground?
- How does basalt act as a carbon sink in ocean?
- How does iron fertilization affect the marine ecosystem?

REFERENCES

Casella, E and Soussana, J.F. (1997). Long-term effects of CO_2 enrichment and temperature increase on the carbon balance of a temperate grass sward. *Journal of Experimental Botany*. **48:** 1309–1321.

Falkowski, P., Scholes, R., Boyle, E., Canadell, J., Canfield, D., Elser, J., Gruber, N., Hibbard, K., Högberg, P., Linder, S., Mackenzie, F., Moore, B., Pedersen, T., Rosenthal, Y., Seitzinger, S., Smetacek, V. and Steffen, W. (2000). The Global Carbon Cycle: A Test of Our Knowledge of Earth as a System. Science (New York, N.Y.). **290**: 291–296.

Idso, K.E. and Idso, S.E. (2006). Global Warming: Carbon Sequestration to Mitigate. Taylor & Francis.

Lal, R., Kimble, J.M., Follett, R.F. and Cole, C.V. (1998). The Potential for U.S. Cropland to Sequester Carbon and Mitigate the Greenhouse Effect. Sleeping Bear Press.

Langdon, C., Takahashi, T., Sweeney, C., Chipman, D., Goddard, J., Marubini, F., Aceves, H., Barnett, H. and Atkinson, M. (2000). Effect of Calcium Carbonate Saturation State on the Calcification Rate of an Experimental Coral Reef. Global Biogeochemical Cycles – GLOBAL BIOGEOCHEM CYCLE. **14:** 639–654.

Chapter 24
Soil Fertility and Nutrient Management

INTRODUCTION

Soil is accumulation of three aspects, viz. minerals, organic materials and living forms. It is storehouse of nutrients and home for organisms. As a medium it provides moisture and nutrients for plant growth. Crop production primarily depends on fertility of soil. Evaluation of soil fertility consists of estimating the nutrient-supplying power of a soil. Proper evolution of soil fertility is required in adopting appropriate measures to ensure good crop production. These methods for evaluating the soil fertility may be biological and chemical. Biological method consists of rising a crop or microbial culture in a field and is useful to measure soil fertility directly but is time consuming. Hence, this method is not well-accepted. On the other hand, chemical process for testing is comparatively rapid. This method for evaluating the soil fertility consist of analyzing sample of the soil for its content of essential plant nutrients. Deficiency of plant nutrients can be compensated by adding suitable fertilizers. The choice of chemical method to estimate nutrient availability in soils involves selection of appropriate extractant along with the extracting procedure which may together remove the nutrient quantities that are highly correlated with the biological index of nutrient availability of group of soil.

HISTORICAL DEVELOPMENTS OF SOIL FERTILITY

The concept of soil fertility to improve crop yields is not a recent approach and is as old as the development of agriculture. In ancient times they had knowledge of applying manures to soil to enhance its fertility. The merits of green manure crops were first recorded by Greek historian Xenophan. Cato (234-149 BC) wrote a practical handbook and recommended intensive cultivation, crop rotations and also classified land based on specific crops. In 1804 Swiss physicist Nicholas Theodore de

S. Mukherjee, *Current Topics in Soil Science*,
https://doi.org/10.1007/978-3-030-92669-4_24

Seussure and in 1834 French chemist Jean Baptiste Boussingault reported that plants need mineral nutrients for growth and development.

SOIL FERTILITY

The interactions of physical, chemical and biological properties in soil control the soil fertility. Soil fertility is defined as the inherent capacity of soil to provide essential nutrients to plants in adequate quantities to maintain the growth and development of specified plant. In brief, growth of crops depend on it. Soil fertility is the key to successful crop production and it is a measure of capacity of soil to supply plant nutrients.

Soil Fertility Factors

The two main factors affecting soil fertility are natural and soil management. Natural factors influence the formation of the soil. It includes parent material, climate and vegetation, topography and age of soil (Table 24.1).

Soil management factors include entire soil management practices depending on soil conditions. The soil management practices indicate addition or removal of plant nutrients. Physical conditions of soil depend on texture and structure of soil. Presence of air and water is very important. Oxygen is essential for root respiration and decomposition of organic matter. Presence of higher content of carbon dioxide in soil restricts the uptake of nutrients. Clayey soil contains more water than sandy soils and, hence, are more productive. Soil compaction decides the aeration status of soil and root penetration and, thus, affects the soil fertility status. Root helps in absorption of water and nutrients which are required for plants. Decomposition of organic matter also depends on soil temperature. Soil microorganisms have a significant role in decomposition; thus, they help to improve the soil fertility. Temperature variations affect the microbial activity. In soils availability of nutrients

Table 24.1 Natural factors of soil fertility

Parent material	Rocks and minerals act as raw materials for the formation of any soils. The property of parent rock controls the property of soil. If the parent material is rich in plant nutrients then the soils will be quite fertile.
Climate and vegetation	Type of vegetation depends on climate especially rainfall and temperature. Temperate soils are not very fertile compared to tropical soils because of the lesser quantity of decomposed organic matter in soil. The tropical soils are more fertile soils due to constant high temperature which helps in faster rate of decomposition of organic matter in the tropics than in temperate regions.
Topography	Generally soils of hilly regions are poor due to erosion and leaching.
Age of soil	Old soils are less fertile due to excessive weathering, leaching.

is influenced by pH. Organic matter increases humus content, hence, availability of higher organic matter helps to increase fertility status.

SOIL PRODUCTIVITY

Soil productivity indicates the ability of a soil for producing plants under a specified system of management. All productive soils are fertile but all fertile soils may not be productive due to adverse climate, salinity, alkalinity etc. To be productive, soil must contain all the essential nutrients required for plants. Availability of balanced proportions of soil nutrients are necessary.

Soil Productivity Factors

The factors affecting soil productivity include physical, chemical and biological conditions of the soil environment. They also include all the practices that affect fertility, water and air relationships and the activity of biological agents microorganisms. Mainly there are two kinds of factors, viz. internal and external. Generally, internal factors are genetic factors which include soil type and texture. External factors include climatic, soil and biotic. Climatic factors include rainfall, solar radiation, presence of atmospheric gasses etc. Soil moisture, soil air, soil temperature are included in soil factors. Physiographic factors and anthropogenic factors are also significant(Fig. 24.1).

Soil fertility, considered as an index of nutrient availability to plants, indicates soil potentiality to produce crops and depends on physical, chemical and biological factors of soil. On the other hand, soil productivity indicates the ability of the soil for crop yield and depends on soil fertility condition along with climatic condition.

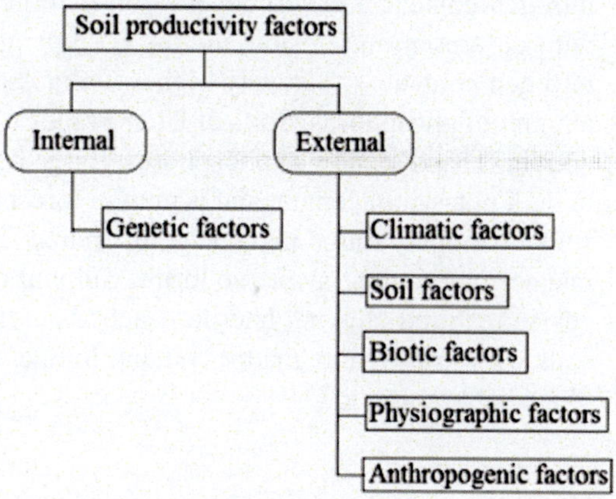

Fig. 24.1 Various soil productivity factors.

It differs with variation of climatic condition and is the function of soil fertility, climate and crop management.

NUTRIENT CYCLE

Soil contains numerous nutrients that are being changed from one form to another by natural processes. Availability of nutrients in soil changes over time by continuous addition and removal from the system and occurs by the influence of geological, chemical and biological processes occurring in soil. Cycling process through which nutrients enter into the soil and out from the soil is known as the nutrient cycle. The nutrient balance indicates the differences between the volume of input and output of constitutes. Input constitutes include mineral fertilizers, organic manures, crop residues, atmospheric deposition, biological nitrogen fixation and sedimentation. Output constitutes include crop removal, run-off and erosion, leaching and gaseous losses. When nutrient additions to soil are higher than their removal then it is called positive nutrient balance which indicates inefficient farming system. On the other hand, negative nutrient balance indicates that unsustainable farming system over long time. Maintaining proper nutrient balance is required to make the system sustainable.

SOIL FERTILITY CONSTRAINTS

Intensive agriculture using various hybrid but without the application of the required amount of nutrients through fertilizers has been leading to the depletion of huge number of nutrients in soil below the critical level over the years which is not good for the health of soil. Productive condition of soil is also hampered. Fertility constraints, the key factors to maintain sustainable agriculture, are nitrogen, phosphorus, sulphur, calcium, magnesium, potassium and micronutrients. The total nitrogen-content in a soil depends on its organic matter content, which decreases with deforestation, erosion and high temperature. In hilly region, North of India, nitrogen content is relatively higher. Indian soils are characterized by overall low amount of phosphorus. Soils of Uttar Pradesh, Uttarakhand, Andhra Pradesh, Bihar and Jharkhand are low in phosphorus. Large percentage of Indian soil have medium to high potassium content and is high in forest hill soil of Uttar Pradesh and alluvial soils of Punjab. Large percentage of Indian soils contains enough of calcium and magnesium in plant-available forms. Sulphur deficiency is generally widespread in coarse textured alluvial, laterite, leached acidic soils, black clayey soils as well as soils with low organic matter content. Saline, alkali soils contain adequate amount of available sulphur.

Effects of Fertilizers on Soil

Intensive use of fertilizer is not only for intensive production but also numerous problems. Excessive use of fertilizer also has negative effect. Experimental investigations along with field study by several workers reveal that the chemical fertilizers have direct impact on the porosity, permeability, water holding capacity, bulk density and compaction of the soil. The soil, treated with organic manures has a notably higher water holding capacity and soil moisture content than those treated with inorganic fertilizers. The faunal activity, enhanced by the organic manure, improves the soil aggregation and porosity by leading to increased infiltration rates (Malik et al., 2014). Application of inorganic fertilizers leads to the formation and accumulation of inorganic salts in the soil, which helps to decrease porosity and subsequently increase compaction in soil. Moreover, it increases bulk density of the soil and makes soil more resistant against penetration of plant roots, thus hindering root development and also plant growth. On the other hand, application of organic manure is to increase water holding capacity of the soil. Long-term application of inorganic fertilizers can bring the changes in nitrogen content, pH and moisture content of soil. Soils, treated with inorganic fertilizers show significantly less microbial populations. Furthermore, long term application of the chemical fertilizers may be extremely harmful to the plant-friendly bacteria groups.

Nitrate pollution in Soil and Groundwater

Nitrate can be absorbed by plants through leaching. Nitrate leaching below the root zone constitutes a potential pollution threat for soil and groundwater bodies. Nitrification and denitrification are the process by which nitrogen dioxide is formed. Nitrate can also be absorbed by plants especially by vegetables and the toxicity also transmitted to human body by consumption. Generally, if the plant is in contact with nitrate for long time, then the chance of accumulation of nitrate will be greater.

Concentration of nitrate in soil profile during leaching and the quantity of water passing through the soil profile are two major factors controlling leaching losses of nitrate. Intensity and amount of rainfall are important in determining the rate and extent of leaching. If the high fertilizer rate is combined with heavy irrigation regimes on coarse-textured soil then the leaching loses of nitrate can be substantial.

Deforestation, extensive cultivation, burning of trees, soil erosion, soil degradation, nutrient losses, excessive fertilizer application etc., have influenced the depletion of soil fertility status. Crop management practices, efficient use of fertilizers and integrated nutrient management practices must be adapted to improve and maintain good soil fertility as well as physical and chemical conditions of soil.

NUTRIENT MANAGEMENT

Soil management is the application of practices to protect soil and enhance its quality and fertility and also helps to improve nutrient cycling by increasing organic matter in soil. It is effective to prevent soil pollution. Nutrient management, a part of soil management, requires to achieve optimum crop yields by applying nutrients in proper quantities. Every crop needs different types of nutrients in different ratios. The principle of nutrient management is to make balance between soil nutrients and crop requirement.

Soil Characteristics

Soil characteristics such as texture, structure, drainage, moisture, pH and temperature affect nutrient uptake. Soil texture is characterized by the proportion of sand, silt and clay in the soil. If clay content increases then the ability to hold nutrients will be also increased. Sandy soils contain maximum porosity, hence, chances of leaching of nutrients is greater. The arrangement of soil particles are defined by soil structure. Good soil structure is represented by significant aggregation and root growth and availability of nutrients depend on it. As soil moisture is essential for root growth, adequate moisture content improve uptake of nutrients. Essential components cannot be lost through denitrification under poor drainage conditions. Growth of plants become affected by temperature and thus nutrient uptake also gets affected. Soil pH affects the availability of most nutrients. At low and high pH conditions, phosphorus is less available than when the pH is around 6.5. Essential nutrient elements have been given in Table 24.2.

Table 24.2 Essential nutrient elements (Baker et al., 1997)

Nutrient	*Category*
Carbon, Hydrogen, Oxygen	Non-fertilizer elements supplied through air, water and soil components
Nitrogen, Phosphorus, Potassium	Macronutrients required by plants in large amounts
Calcium, Magnesium, Sulphur	Secondary nutrients required by plants in moderate amounts
Boron, Chlorine, Copper, Iron, Molybdenum, Zinc	Micronutrients required by plants in small amounts

Integrated Nutrient Management (INM)

Integrated Nutrient Management is a good approach to maintain soil fertility and supply adequate nutrients to restore degraded soil. Applications of chemical fertilizers along with organic manures are included in this approach. Basic concept of the principles of INM is to integrate all sources of plant nutrients and also all improved crop production technologies into a productive agricultural system (Roy, 1986). Its approach includes the requirements of nutrients with appropriate proportions, best timing for applying the nutrients and how best these forms should be integrated to obtain the highest production in an environment friendly manner. INM integrated technologies are site specific rather than nutrient application because nutrient application does not depend on crop type but also on climate, socio-economic conditions and physical, chemical and biological condition of soil. Components of INM include primarily chemical fertilizers and biofertilizers. Biofertilizers include biological nitrogen fixation and biosolids. Animal manures and composts, byproducts and wastes, plant residue etc. are included into biosolids.

Plant Nutrients

Plant nutrients are the elements that are essential for plant growth and are taken from the soil. Water can be another source for plant nutrients through irrigation, flood or groundwater.

Nitrogen, phosphorus and potassium are the primary nutrients which are consumed in relatively large amounts. Three secondary nutrients include calcium, magnesium and sulphur. These are essential for plant growth but taken up in smaller quantities. Micronutrients are also required as essential constituents and are taken in very small quantities for plant or animal metabolism. Integrated Plant Nutrition Systems is used to maintain soil fertility and plant nutrient supply to achieve good amount of crop production. It is done by optimizing the benefits of plant nutrients. The main objectives are to enhance soil productivity by using of mineral fertilizers combined with plant nutrients and to improve the stock of plant nutrients in soil.

THINK FOR A WHILE

- Is there any difference between soil fertility and soil productivity?
- How the plant nutrient balance system is influenced by soil fertility?
- What is optimum fertilizer rate?
- What is the role of earthworms in maintaining soil fertility?
- Which nutrient is used in the largest amount as fertilizer?
- How nitrogen pollution is reduced in N management?

- What kind of tillage is practiced in K management and why?
- What are the advantages of precision farming?
- What is the best approach to quantify the release of phosphorus fertilizers in soils?

REFERENCES

Baker, R.D., Shane, T. and Flynn, R. (1997). Soil Analysis: A Key to Soil Nutrient Management Guide A-137. Cooperative Extension Service, College of Agriculture and Home Economics, pp. 1-4.

Malik, S.S., Chauhan, R.C., Laura, J.S., Kapoor, T., Abhilashi, R. and Sharma, N. (2014). Influence of Organic and Synthetic Fertilizers on Soil Physical Properties. *International Journal of Current Microbiological Application Science*, **3(8):** 802-810.

Roy, R.N. (1986). FAO, fertilizer programme with special reference to integrated plant nutrition system. Fertilizer in Agricultural Development. pp. 149-169. Krishak Bharati Cooperative Limited, New Delhi, India.

Chapter 25
Soil Pollution

INTRODUCTION

Soil environment is the function of several interrelated physical, chemical and biological parameters. Gradually increasing industrialization along with agriculture advancement by using various chemical substances leads to undesirable variation in soil components. Not only soil contents but also soil pH and moisture can be adversely affected. Applications of these chemicals also destroy the soil microbial activity. Human survival is dependent on health of soil. Undoubtedly soil pollution is a serious environmental issue. For example, soil containing high concentration of benzene, increases the risk of contracting leukaemia. Besides, the chemical test, soil pollution is indicated by discolouration of soil. Thus, it is essential to estimate the toxicity of containing compounds. It may be either due to man-induced sources or by natural sources. In some cases both are responsible.

The main reason of contamination is the presence of man-made waste. The waste products, full of chemical compounds are not originally found in nature and lead to soil pollution. The significant causes of soil pollution are improper use of pesticides and excessive industrial activity. Inefficient disposal of industrial waste also plays a key role in soil pollution. Industrial wastes have different amount of hazardous chemicals. By the process of contamination high toxic affects the soil layer strength in the top soil and leads to soil pollution. Thus, fertility and biological activity of the soil are gradually reduced. In addition, hazardous effect also leads to ecological imbalances which troubles crop production. It may cause severe health problem. Large scale deforestation along with improper rotation of crops is the causes of soil erosion.

S. Mukherjee, *Current Topics in Soil Science*,
https://doi.org/10.1007/978-3-030-92669-4_25

SOIL HEALTH

According to Doran and Zeiss (2000) soil health indicates its capacity of a specific type of soil to sustain plant and animal productivity.

Following factors are accepted globally as the soil quality indicators:

- Permeability of the soil.
- The degree to which soil can assimilate organic wastes and absorb or neutralize different types of contaminants to protect our ecosystem.

The above factors are determined by a number of physical, chemical and biological parameters. It has been suggested that soil texture, structure, strength, water holding capacity are physical parameters of soil health and pH, electrolytic conductivity, salinity, and soil organic matter content are chemical parameters. Biological parameters are determined by presence of microbial activity along with abundance of organisms, their diversity, food web structure, and community stability.

POLLUTANTS

Soil pollutant includes organic and inorganic naturally-occurring or man-made chemicals. Some of the most hazardous soil pollutants are xenobiotics, i.e. those substances that are synthesized by human beings. Several xenobiotics are known to be carcinogens.

Heavy Metals

Presence of high concentrations of heavy metals in soils can cause it to become highly toxic for human beings. Toxic metals like arsenic, antimony, cadmium, thallium, mercury, lead, zinc, nickel, selenium, beryllium, chromium and copper can be classified as soil pollutants. These metals can originate from several sources such as mining activities, agricultural activities, electronic waste (e-waste), and medical waste.

Polycyclic Aromatic Hydrocarbons

Polycyclic aromatic hydrocarbons are organic compounds that contain carbon and hydrogen atoms and more than one aromatic ring in their chemical structures. Naphthalene, anthracene and phenalene are common examples of Polycyclic Aromatic Hydrocarbons. Hydrocarbon contaminated soil can impose worse effects on forest health and agriculture products. Polycyclic aromatic hydrocarbons are treated as carcinogenic. Soil pollution due to Polycyclic Aromatic Hydrocarbons can be sourced to coke processing, vehicle emissions etc.

Industrial Waste

The discharge of industrial waste into soils is the primary cause of soil pollution. Chlorinated industrial solvents, plasticizers, polychlorinated biphenyls are examples of industrial waste. Dioxins also represent industrial waste produced from the manufacture of pesticides.

Pesticides

Pesticides are generally used to inhibit the growth of pests. Common types of pesticides used in agriculture include herbicides, insecticides and fungicides. Pesticides pose several health risks for human beings. Examples of health hazards related to pesticides include diseases of the central nervous system, immune system, cancer and birth defects. Soil contaminants, found in pesticides are listed in Table 25.1.

Mechanism of waste-soil interaction can be categorized as physical, chemical and biological. Several processes are involved within this mechanism (Table 25.2).

Nature of waste water in soil is influenced by the pore spaces within soil and characteristics of mineral particles of soil. The residence time of waste constituents increase rapidly with water storage capacity of the soil and thus it promotes the interaction of waste with soil. Retention and transmission of dissolved ions in waste are governed by cation exchange capacity of soil. The dissolved ions in waste waters are retained in soil primarily through adsorption and chemical precipitation. Specific

Table 25.1 Some important soil contaminants found in pesticides

Herbicides	• Triazines • Carbamates • Amides • Phenoxyalkyl acids • Aliphatic acids	Used to kill/control weeds and other unwanted plants.
Insecticides	• Organophosphates • Chlorinated hydrocarbons • Arsenic-containing compounds • Pyrethrum	Used to kill insects.
Fungicides	• Mercury-containing compounds • Thiocarbamates • Copper sulfate	Used to kill parasitic fungi or inhibit their growth.

Table 25.2 Mechanism of waste-soil interaction

Interaction mechanism		
Physical	*Chemical*	*Biological*
• Filtration • Dilution	• Adsorption • Precipitation • Inorganic • Transformation	• Assimilation • Decomposition

adsorption occurs when soils retain certain cations and anions which are not exchangeable. Thus, some heavy metal cations such as Zn, Cu, Ni, Cd and Pb can be incorporated into soil. Biological transformation of wastes in soil occurs primarily in the root zone. The diversity of organisms in waste waters enhances soil's ability to degrade a wide variety of organic substances. In soil, amount of oxygen determines the rate of organic matter decomposes. Initially the organisms digest the easily-decomposable materials. In the second stage of biological transformation, microbial tissues from organisms die and some resistant compounds are partially degraded. Thus, humus is formed and degraded by highly specialized organisms during the final stage of biological transformation.

CAUSES OF SOIL POLLUTION

Industries (including agriculture) are believed to be the leading causes of soil pollution due to disposal of the toxic wastes in unorganized way. Chemical fertilizers and pesticides easily contaminate with soil. Extensive use of those used for the growth and maintenance of crops make severe consequence. Leaks in pipes can cause fuel spills during petroleum or diesel transportation. These fuels contain toxic hydrocarbons which contaminate with soil and can cause soil pollution. Soil pollution can be classified into two categories, viz. naturally caused soil pollution and anthropogenic (caused by human activity) soil pollution.

Natural Pollution of Soil

Natural pollution of soil can occur due to differential deposition of soil by the atmosphere. It indicates some extremely rare processes where some pollutants are naturally accumulated in soils. Soil pollution also can occur through transportation of soil pollutants with precipitation water. These types of contaminants can be naturally produced in the soil under the effect of certain environmental conditions. Perchlorates are example of such rare condition. It can be formed in soils.

Anthropogenic Soil Pollution

In maximum cases soil pollution is anthropogenic in nature. Number of human activities can lead to soil contamination. Some such processes are highlighted as follows:

- Metal casting factories discharge metallic effluent which can cause the dispersion of metallic contaminants into the nearby soils.
- Underground mining activities can cause contamination of soil with heavy metals.
- Improper disposal of highly toxic industrial waste can severely pollute the soil. For example, seepage of the waste into the soil can pollute soil and groundwater.
- Contamination between soil and hydrocarbons can occur due to spillage of petrol and diesel during transportation.
- In urbanized areas unorganized way of sewage disposal can contaminate soil.
- Severe soil pollution result by excessive use of chemical pesticides as they contain several hazardous substances.
- Nuclear waste and e-waste can pollute soil as well as environment.
- Demolition of old buildings can lead to the contamination of nearby soil with asbestos.

ROLE OF PESTICIDES

Pesticides are applied to control pests in soil. Inappropriate pesticides application lead to contamination of toxic contents into soil. The effectiveness of pesticides depends largely on the duration for which pesticides remain in soil. DDT has a half-life of three years in cultivated soils, while organophosphate insecticides persist only for a few days or months. Pesticides can be grouped into three types, viz. persistent, non-persistent and moderately persistent. Persistent indicates half-life greater than 100 days, non-persistent shows half-life less than 30 days and moderately persistent indicates half-life greater than 30 days but less than 100 days.

The chlorinated hydrocarbons persist longer in soils having large amount of organic matter, although more amount of the chemical must be applied to these soils to kill pests. There are three types of pesticides (Fig. 25.1). Herbicides applied to soils may not persist at all or may persist for very short duration, depending on the

Fig. 25.1 Types of pesticides.

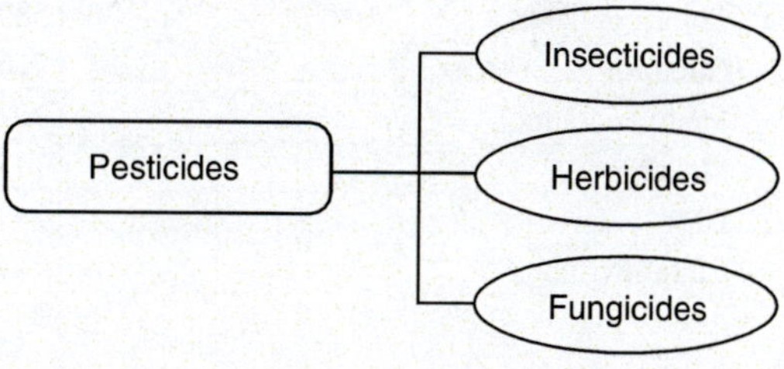

individual compound. Pesticides can be removed through evaporation, vapourization, leaching, plant uptake and chemical decomposition. When pesticide enters the soil some of it is absorbed, some gets dissolved and mixed with soil water. Due to excess water by rain or irrigation the absorbed pesticides molecules may be desorbed from soil. Partition coefficient value is defined as the ratio of pesticide concentration in the adsorbed state and the solution-phase. If pesticides have smaller partition coefficient, then the concentration of pesticides in solution is higher. The partition coefficient is determined by solubility and melting point of a pesticide. Partition coefficient with several pesticides component is given in Table 25.3.

Activities of soil microbial may be affected by pesticides because these are chemicals that have deleterious effect and can be metabolized. Pesticides decrease biodiversity in the soil. Pesticides not only kill pests but also deteriorate soil biology and soil quality.

ROLE OF TOXIC ELEMENTS

The industrial and sewage wastes, and fertilizers are the main sources of toxic elements. These can be air borne and can easily mix with soil through native rocks and mining operations. Toxic elements interact with both organic and inorganic matter. Toxic elements become available to vegetation and consequently it can easily reach

Table 25.3 Partition coefficients of some pesticides (Rao et al., 2008)

Pesticides	*Partition coefficient*
Carbofuran	29
Propham	51
2. 4-D	32
Chloramben	13
Bromacil	72
Prometone	300
Dichlobenil	164
Propazine	207
Carbaryl	229
Monolinuron	237
Diuron	389
Cynazine	190
Chlorpropham	589
Parathion	7161
Methyl parathion	7079
Malathion	1778
DDT	243000
Chloropyriphos	13490

in human and animal body through consumption of those vegetables. Sometimes crop plants exhibits some unusual characteristics features in response to the presence of certain toxic elements. Crop species differ in extracting toxic elements from soil. Soil contamination can be prevented by eliminating toxic content or by managing the soil-crop system in a manner that leads to the prevention of recycling of toxic content. Soil-crop management practices help to keep the chemicals in soil and also reduce their uptake by plant. If the soil pH can be maintained neutral or above, then most of the toxic elements are rendered less mobile and less available. On the other hand, the drainage of wet soils helps in the formation of oxidized forms of several toxic elements, which are generally less available for plant uptake. Lead, nickel, mercury, cadmium, chromium, selenium and arsenic are examples of toxic elements (Fig. 25.2 and Table 25.4).

ROLE OF RADIONUCLIDES

Gamma rays, beta rays and alpha particles can be radiated from radioactive elements. Interaction of living body with radiations brings about a number of changes such as modification of cells consequently persons suffer nausea, diarrhea, vomiting, leukemia. Radioactive contamination might happen through several sources. Waste product and effluents of nuclear reactor are the principal source of nuclear waste which leads to worldwide toxic waste contamination. Contaminated soils and plants are important links in the movement of radioneuclides through food chain. The nuclear accident at the Chernobyl power plant on 26th April, 1986 in the erstwhile U.S.S.R. demonstrated the vulnerable condition of agriculture soils to radioactive contamination from atmospheric fallout. Radionuclides are formed as a result of application of atomic energy. Some of the radionuclides have the important significance in food chain such as $^{89-90}Sr$, ^{137}Cs and ^{131}I. These radioactive elements are being consumed by plants. The uptake of radionuclides by plants depends on the solubility of fallout debris, chemical activity of those nuclides in soil and the absorption power of that particular plant root. According to Nishita and Taylor (1964), the relative order of magnitude of plant uptake of fission product is $^{89-90}Sr > ^{131}I > ^{140}Ba > ^{137}Cs$. It has been noted that the absolute magnitude of uptake depends on the soil management practices such as fertilization, cultivation, application of pesticides etc. Although it is true that transfer factors of radioactive elements such as Cs, Sr, Mn and Zn etc. from soil to plant is generally very low.

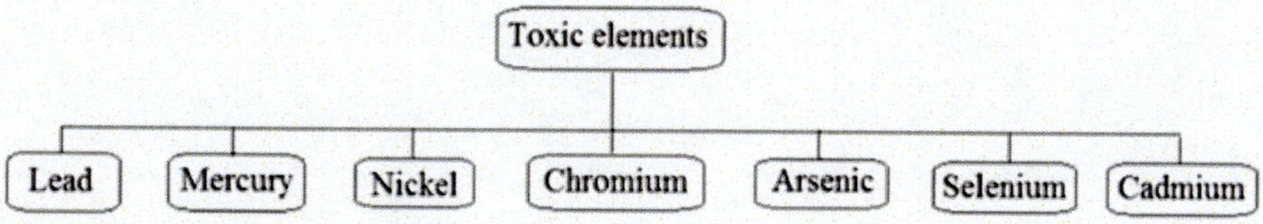

Fig. 25.2 Varieties of toxic elements.

Table 25.4 Description of toxic elements

Elements	*Description*
Lead	The accumulation of lead in soil is primarily a function of the rate of deposition from atmosphere. Sewage and industrial waste also play an important role in contribution of lead. Consequence of lead in soil is affected by exchange adsorption at mineral interfaces and precipitation of solid phases. Lead is easily converted to sulphate at the soil surface. Lead sulphate is less soluble and could leach through the soil. If soil pH is 5 or more along with approx. 5% organic matter then lead is retained in the upper 2-5 cm of undisturbed soil. Lead may mobilize if lead bearing soil particles runoff to surface waters during heavy rains. Smaller lead containing soil particles may be mobilized due to effect of prevailing winds. Under natural conditions the downward movement of lead by influence of leaching process is very low. High concentration of lead is required to produce toxic response. Generally, in response of the hydrolysis and polymerization process, lead is fixed in soil. High cation exchange capacity, alkaline pH, high organic matter of soil reduces lead-uptake by plants.
Nickel	Highest Ni concentration in soil is found near industries where Ni is extracted from ore. Nickel pollution may affect microbial community development and its activity in soil which subsequently affect the fertility of soil. Nickel is the most stable state over a wide range of pH and redox conditions found in surface soils. Mobility of Ni in soils increases as pH and cation exchange capacity decreases. Range of 50-100 mg/g of Ni content is indication of its toxicity in plants. If industrial and municipal wastes with high Ni concentration are applied to agriculture land then possibility of Ni-toxicity increases.
Mercury	Mercurial fungicides (Ceresan M.) constitute significant source of Hg for agriculture. Behaviour of mercury in soil is controlled primarily by adsorption and desorption processes. Hg content of the aboveground plant parts is generally low and this element becomes more hazardous for aquatic food chain than agriculture.
Cadmium	In and around mining areas cadmium level is found very high. Cadmium is very toxic element to a number of plant species. It adversely affects microbial activity involved in nutrient cycling and it also hampers in maintenance of soil fertility. Consumption of cadmium by plants is controlled by its concentration in soil. According to plant species accumulation of cadmium also varies. Cd has tendency to concentrate in leafy vegetable tissues. Presence of heavy metals, soil pH influence uptake of Cd by plants. Increase in soil pH reduces the soil availability of Cd to crop plants. Incidence of itai-itai disease in Japan occurred due to presence of high Cd content in rice.

(continued)

Table 25.4 (continued)

Elements	*Description*	
Chromium	In soil chromium is found in two main oxidation forms, viz. Cr(III) and Cr(VI) having distinctly contrasting properties	
	Cr(III)	*Cr(VI)*
	Cr(III) is apparently harmless to living organisms at reasonable concentrations	Cr(VI) is harmful to living organisms due to its toxicity
	Risk of its leaching is negligible because of its immobility in soil	Present in forms of chromates and dichromates and both are mobile in soil
	Roots are the main obstruction for entry of Cr into plants. Generally it is fixed on roots and leaching chances are very low. Thus, plants do not accumulate higher concentration of Cr from soil.	
Selenium	Selenium is not essential for plants and is present as elemental Se, selenites bound to hydrous iron oxides, heavy metal selenides etc. The Se-forms, when applied to soil, are not easily available to plants. Thus, chances of occurrence of selenium hazards from selenium rich waste water or sludge are little.	
Arsenic	Enrichment of arsenic in soil occurs due to weathering of arsenic bearing rocks or use of arsenic contaminated groundwater for irrigation. Anthropogenic activities like mining and ore exploration are also responsible. Soils and overlying sulphide ore deposits have high arsenic contents. Arsenic can be released through oxidation of arsenic rich sulphide compounds. Arsenic retention soils are governed largely by the adsorption process rather than the precipitation of arsenic compounds. Toxicity of arsenic compounds in soil is determined by oxidation state and its presence in the organic/inorganic combination.	

CONSEQUENCES OF SOIL POLLUTION

A broad spectrum of negative consequences of soil pollution affects plants, animals and human beings. Children are more susceptible to such diseases caused by soil pollution. Polluted soil poses a greater threat to the whole ecosystem. Some important effects of soil pollution are highlighted as follows:

Effects on plants and animals

Soil pollution is accompanied by decreasing nutrients. Soil fertility is also hampered due to presence of toxic elements. Soil contamination often increases the salinity of the soil. All the pollutants, accumulated by plants are passed up the food chain.

Effects on human beings

The contaminants found in polluted soil can enter human bodies nose, mouth, or skin as the soil contaminants can exist in all three phases, viz. solid, liquid and gaseous. Exposure to such soils can cause a variety of short-term health problems but long-term exposure to contaminated soil can lead to severe health hazards and damage to vital organs. Petroleum hydrocarbons and industrial solvents have been linked to congenital disorders in humans. The short term effects of human exposure to polluted soil include headaches, nausea, pain in chest, skin diseases etc. On the other hand long term effects include permanent damage to the nervous system, cancer etc.

Effects on the Ecosystem

Soil pollution can be a direct contributor to air and water pollution as the volatile contaminants in the soil can be carried into the atmosphere by wind or can seep into underground water reserves. It can also contribute to acid rain. Acidic soils are inhospitable to several microorganisms which are required to improve soil texture and help in the decomposition of organic matter. Thus, the soil quality and texture become affected. Along with decreasing soil quality crop yield is greatly affected due to contamination with heavy metals.

SOIL POLLUTION CONTROL

Contaminated soil is one of the environmental problems and has an ecological impact. Soil can be remediated by using appropriate technologies. Many crucial changes must be brought easily in order to control soil contamination and soil pollution.

To control soil pollution following strategies have been followed:

- Use of soil-friendly agricultural process
- Planting of trees to check soil erosion in coastal areas and mountainous regions
- Controlled grazing system
- Formulation and implementation of stringent pollution control legislation
- Improved sewage system in urban areas
- Proper disposal of toxic wastes including industrial, household and agriculture
- Recycling and recovery of nontoxic materials
- Reduction of waste generation by improving reuse process
- Use of chemical fertilizers
- Minimize use of pesticides

Several technologies have been developed for soil remediation. Some important strategies, followed for the decontamination are given in Table 25.5.

To manage the consequence due to hydrocarbon contaminated soil well established monitoring systems are required. Remote sensing is one of the most significant for such monitoring systems and is highly used for those areas with number of difficulties to access. To curb the negative effects of soil pollution some environmental remediation processes are taken such as decontamination and excavation of wastes. The contaminated soil is required to be excavated and transported to a remote disposal site. Thermal remediation of contaminated soil involves heating of the soil to reduce the volatile toxic pollutants through vapourization.

Reducing Heavy Metal Pollution

Heavy metals play an important role in soil contamination. Quality of soil can be severely damaged through the addition of inorganic toxins. Prevention of contamination can be achieved by eliminating toxicity in soil. Arsenic uptake by different crops is listed in Table 25.6.

Crop management practices can reduce the cycling of toxic inorganic chemicals. By following these practices it is possible to reduce their uptake by plants. If the soil pH can be maintained neutrally then most of toxic elements are rendered less mobile. Similarly, drainage of wet soils helps in the formation of oxidized forms of several toxic elements, which are less soluble and less available for plant uptake. Availability of toxic cations can be reduced through heavy phosphate application. Hence, some crop species have the ability to extract toxic elements from soil and this property can be used to decontaminate the soils.

Table 25.5 Technologies for soil remediation

Excavation	Excavation of polluted soils to remote, uninhabited locations
Thermal remediation	Extraction of pollutants vapour extraction
Mycoremediation	Involves use of fungi for accumulation of heavy metal contaminants
Phytoremediation	Involves use of microorganisms and plants for decontamination of soil

Table 25.6 Arsenic uptake by different crops (Das, 2007)

Type of crops	*Leaf (mg/kg)*	*Stem (mg/kg)*	*Root (mg/kg)*
Jute	3.69	9.3	11.02
Potato	5.2	7.8	12.2
Mustard	6.96	7	7.23
Rice (summer season)	8.42	7.32	8.3
Rice (wet season)	5.12	6.3	8.23
Groundnut	2.12	2.43	3.1
Wheat	5.80	6.93	14
Maize	3.8	6.19	6.32
Green gram	5.28	5.7	6

ROLE OF CLAY MINERALS IN PROTECTION OF SOIL ENVIRONMENT

Clay minerals and their derivatives may be used as highly effective adsorbents for a wide range of inorganic and organic contaminants. The clay minerals with 2:1-structural types are good adsorbents for highly polar cationic contaminants due to their small particle-size, very large specific surface areas and their hydrophilic, negatively-charged surfaces. Their derivatives can adsorb anionic pesticides by ion-exchange reactions. The clay minerals belong to the phyllosilicate subclass.

K^+, Na^+ and Mg^{2+} are the most common interlayer cations in illite, smectite and vermaculite respectively. These cations may be readily replaced when the mineral surface is in contact with a liquid containing different types of cations like Cu^{2+}, Pb^{2+}, Zn^{2+}, Cd^{2+}, Mn^{2+} etc. Cation exchange capacity for illite, smectite and vermiculite are 10-40 meq/100 gm, 40-150 meq/100 gm and 33-174 meq/100 gm, respectively (Silva et al., 2018).

Natural clays, especially those having t–o–t–c structures, are effective adsorbents for cationic and highly-polar pesticides owing to their internal structures and also for their physical and chemical properties. They can mix well with the pesticide particles due to their very fine grain size. Clays are very good adsorbents for cationic and highly-polar pesticides because of having the hydrophilic, negative character of their surfaces. They have very thin, flaky grain shape which maximizes surface area of the grains and enhance the surface reactions. Bentonite, montmorillonite are the clay minerals of smectite group. These are helpful for adsorbing the basic pesticides.

THINK FOR A WHILE

- What are the significant processes that cause soil pollution?
- What is the role of lead and mercury in contamination that pollute soils?

- What is the role of waste management to avoid soil pollution?
- How the internal structure of clay is related with their adsorption property?
- What are the key causes of soil pollution?
- What are the environmental remediation processes to reduce the negative effects of soil pollution?
- What are the effects of soil pollution on human health?

REFERENCES

Das, D.K. (2007). Effects of arsenic-contaminated irrigation water, zinc and organic matter on the mobilization of arsenic in soils in relation to rice (*Oryza sativa* L.). *Trace Metals and Other Contaminants in the Environment*, **9:** 331–354.

Doran, J.W. and Zeiss, M.R. (2000). Soil health and sustainability: Managing the biotic component of soil quality. *Applied Soil Ecology*, **15:** 3–11.

Nishita, H. and Taylor, P. (1964). Influence of stable Sr and Ca on 90Sr and 45Ca in soil and clay minerals. *Soil Science* **98:** 123–142.

Rao, P.C.S., Mansell, R.S., Baldwin, L.B. and Laurent, M.F. (2008). Pesticides and their behavior in soils and water. http://pmep.cce.cornell.edu/facts-slides-self/facts/gen-pubre-soil-water.html. Accessed 24 January 2009.

Silva, V., Montanarella, L., Jones, A., Fernandez-Ugalde, O., Mol, H.G.J., Ritsema, C.J., et al., (2018). Distribution of glyphosate and aminomethylphosphonic acid (AMPA) in agricultural topsoils of the European Union. *Sci. Total Environ.* **621:** 1352–1359.

Index

S. Mukherjee, *Current Topics in Soil Science*,
https://doi.org/10.1007/978-3-030-92669-4